SpringerBriefs in Computer Science

SpringerBriefs present concise summaries of cutting-edge research and practical applications across a wide spectrum of fields. Featuring compact volumes of 50 to 125 pages, the series covers a range of content from professional to academic.

Typical topics might include:

- A timely report of state-of-the art analytical techniques
- A bridge between new research results, as published in journal articles, and a contextual literature review
- A snapshot of a hot or emerging topic
- An in-depth case study or clinical example
- A presentation of core concepts that students must understand in order to make independent contributions

Briefs allow authors to present their ideas and readers to absorb them with minimal time investment. Briefs will be published as part of Springer's eBook collection, with millions of users worldwide. In addition, Briefs will be available for individual print and electronic purchase. Briefs are characterized by fast, global electronic dissemination, standard publishing contracts, easy-to-use manuscript preparation and formatting guidelines, and expedited production schedules. We aim for publication 8–12 weeks after acceptance. Both solicited and unsolicited manuscripts are considered for publication in this series.

**Indexing: This series is indexed in Scopus, Ei-Compendex, and zbMATH **

Liting Wang · Xiaoming Tao · Lu Sun ·
Wentao Shen

Multi-spectral and Intelligent Sensing

 Springer

Liting Wang
Department of Electrical Engineering
Tsinghua University
Beijing, China

Xiaoming Tao
Department of Electrical Engineering
Tsinghua University
Beijing, China

Lu Sun
Department of Mathematical Sciences
Tsinghua University
Beijing, China

Wentao Shen
Department of Electrical Engineering
Tsinghua University
Beijing, China

ISSN 2191-5768   ISSN 2191-5776 (electronic)
SpringerBriefs in Computer Science
ISBN 978-981-97-4549-4   ISBN 978-981-97-4550-0 (eBook)
https://doi.org/10.1007/978-981-97-4550-0

This Springer imprint is published by the registered company Springer Nature Singapore Pte Ltd.
The registered company address is: 152 Beach Road, #21-01/04 Gateway East, Singapore 189721,
Singapore

If disposing of this product, please recycle the paper.

Contents

Chapter 1
Multi-spectral Sensors and Applications in Various Domains

Abstract This chapter introduces multispectral sensing technology and its applications in agriculture, environmental monitoring, medical imaging and other fields. Multispectral sensing technology detects and analyzes target substances by capturing spectral band information at different wavelengths, which can reveal more material properties of target objects than traditional imaging technologies. Modern multispectral sensors are rapidly developing towards higher spectral resolution, more powerful data processing capabilities, lower costs, and more compact integrated designs. At the same time, with miniaturization, their application scope extends to portable devices, drones and other new platforms. Combining machine learning and artificial intelligence algorithms, data analysis and application are more accurate and efficient.

Keywords Multispectral sensors · Intelligent sensing · Infrared spectroscopy · Terahertz spectroscopy · Safe rural areas

1.1 Overview

1.1.1 Introduction to Multi-spectral Sensing Technologies

Multispectral sensing technology is an advanced technology that detects and analyzes target substances by capturing spectral band information at different wavelengths. Compared with traditional panchromatic imaging or single-wavelength imaging technology, multispectral sensors can acquire light signals at a specific range of wavelengths, thereby revealing more material properties of the target object, such as chemical composition, temperature, humidity and other important environmental characteristics. This technology has extremely important applications in many fields, such as precision agriculture [1] using it to monitor crop growth and soil conditions, environmental monitoring [2] using it to observe natural ecological changes and pollution, and medical imaging [3] which is used to diagnose diseases and monitor conditions.

The structure of a multispectral sensor generally contains several spectral channels, each of which is precisely tuned to a different wavelength range. By comparing and analyzing spectral data in these different frequency bands, experts can obtain the unique spectral characteristics of the target substance. These characteristics can be used for accurate identification, classification and even complex quantitative analysis of objects. For example, in agriculture, moisture content and health status can be determined by analyzing specific wavelengths of light reflected by crops; in environmental monitoring, multispectral sensors can reveal changes in vegetation cover and the spread of pollutants.

With the advancement of science and technology and the increasing demand for high-precision data, modern multispectral sensors are rapidly evolving towards higher spectral resolution, more powerful data processing capabilities, lower costs and more compact integrated designs. At the same time, the miniaturization of multispectral sensing technology has expanded its application scope to emerging platforms such as portable devices and drones, thereby providing new possibilities for remote sensing and field data collection. In addition, with the integration of machine learning and artificial intelligence algorithms, data analysis and application are more accurate and efficient, making multispectral imaging not only limited to researchers in professional fields, but its ability to assist decision-making and analysis has increasingly become a necessary tool in all walks of life.

Multispectral sensing technology is in a rapid development stage, and its potential and application prospects are exciting. From agricultural management to environmental protection, from medical diagnosis to resource exploration, multispectral sensing is gradually becoming one of the indispensable key technologies. In the future, it is expected to further promote the in-depth integration of scientific research and industry practice, providing us with a more comprehensive and in-depth understanding of the world.

1.1.2   Evolution and Trends of Multi-spectral Sensors

Multispectral sensor technology has come a long way since it was first used in aerospace remote sensing in the 1960s. Due to limitations in technology and cost, early multispectral sensors had relatively low number of bands and low resolution. However, with the advancement of semiconductor technology and digital signal processing technology, modern multispectral sensors have been able to provide more wavelength bands, higher resolution, and better sensitivity [4].

In recent years, with the innovation of microelectronics and photonics technology, multispectral sensors have begun to develop towards miniaturization, low cost, and high integration. In addition, with the rise of drones and microsatellite platforms, the demand for lightweight and low-power multispectral sensors has surged, further promoting innovation in sensor technology.

Currently, an important trend in multispectral sensors is the evolution towards hyperspectral imaging technology. Hyperspectral sensors can provide hundreds of subdivided continuous spectral bands over a wider wavelength range, which makes it possible to accurately identify and analyze material components [5].

At the application level, multispectral sensors are being integrated into a wider range of platforms and systems, such as drones, robots, mobile devices, and wearables. Through built-in multispectral sensors, these devices can perform real-time monitoring and analysis in the wild or in daily environments, greatly expanding the application fields of multispectral technology.

In the future, with the combination of artificial intelligence and machine learning technology, the data processing and interpretation capabilities of multispectral sensors are expected to be further improved, which will promote the intelligent application of multispectral sensing technology in various industries [6].

1.1.3 Basic Principles of Multispectral Imaging

The core of a multispectral imaging system is its ability to simultaneously acquire images of the same scene in different spectral bands. These bands typically cover different ranges of the electromagnetic spectrum, from visible light to infrared. Each band corresponds to different spectral information, which is extremely valuable in analyzing vegetation coverage, water quality analysis, rock mineral composition identification.

Multispectral sensors are usually composed of multiple independent sensor arrays, each array is only sensitive to a specific spectral band. These arrays are capable of simultaneously capturing optical radiation from a target scene to produce multiple independent images. The light entering the sensor is separated into different wavelength bands through optical filters or gratings, and then the light in each band is captured by a specialized detector array [7].

The design and manufacturing of multispectral sensors is a complex process that requires precise control of the wavelength range and bandwidth of each band. The choice of wavelength depends on the application requirements and the spectral characteristics of the target object. For example, vegetation reflects light in the green band and absorbs red and blue, and infrared bands are often used to assess the health of vegetation.

In the fields of mathematics and computing, the processing of multispectral imaging data involves complex algorithms, including spectral analysis, feature extraction, statistical classification, and pattern recognition. The purpose of the algorithm is to extract the most useful information from multi-band data and transform this information into a form that is easy for users to understand. Some advanced multispectral imaging systems also combine machine learning and artificial intelligence technologies to automatically identify and classify patterns in images [8].

With advances in optical and electronic technology, multispectral sensors have become more compact in size and have higher spectral and spatial resolution. The new sensor design also allows for faster image acquisition rates and lower noise levels, improving imaging quality and system utility.

1.1.4   Research and Technical Challenges

Although multispectral imaging technology has made great progress, its research and application still face many technical challenges. The first is to improve the spectral resolution and sensitivity of the sensor. Current multispectral sensors still have limitations in the coverage and resolution of specific wavebands, which is particularly difficult to accurately identify substances with similar spectral characteristics [9].

In addition, improving the spatial resolution of imaging systems is also a challenge, especially on drones and satellite platforms that need to be miniaturized and lightweight. In order to achieve higher quality imaging with limited load capacity and volume, new lightweight materials and more efficient imaging technologies need to be developed.

The processing and analysis of big data is also a key challenge. The large amount of data generated by multispectral imaging systems requires the use of advanced algorithms for efficient preprocessing, dimensionality reduction, and feature extraction. These algorithms are designed to increase processing speed and reduce storage space requirements while maintaining the accuracy of analytical results.

Calibration and validation of multispectral imaging systems are equally important. The calibration process ensures that the image data truly reflects the spectral properties of the target. Therefore, precise calibration needs to be performed regularly, and more advanced calibration techniques need to be developed to ensure data reliability [10].

As technology advances, future research will also focus on the fusion of multispectral data with other sensor data (such as thermal imaging, lidar). Improvements in data fusion technology are expected to provide new possibilities for more comprehensive environmental perception and target recognition.

1.2   Spectral Extension—The Invisible Light

1.2.1   Infrared (IR) Spectroscopy Technology

Infrared spectroscopy is a non-destructive analytical method that can obtain information on the chemical composition of a sample without direct contact with it. The technology is based on a core principle: different molecules absorb energy in the spectrum at specific infrared wavelengths, creating a unique infrared spectral fingerprint. Therefore, infrared spectroscopy technology has been widely used in the analysis of organic and inorganic compounds, biomedicine, environmental monitoring, artwork identification and other fields [10].

Infrared spectroscopy technology is mainly divided into three sub-regions: near infrared (NIR), mid infrared (MIR) and far infrared (FIR). The near-infrared region is mainly used to analyze components such as water, fat, and protein. The mid-infrared region can be used to obtain more detailed molecular vibration and rotation information. The far-infrared region is usually used to study lattice vibration or low-frequency molecular vibration [11].

A key challenge in infrared spectroscopy technology is improving sensitivity and resolution. In recent years, with the development of detector technology and optical components, advanced technologies including Fourier transform infrared spectroscopy (FTIR) have been able to provide higher spectral resolution and wider dynamic range [12].

In addition, the combination of infrared spectroscopy technology with other analytical methods, such as Raman spectroscopy, mass spectrometry, and multispectral imaging, also brings new opportunities for material analysis. These multimodal analysis technologies allow researchers to comprehensively analyze samples from different perspectives and obtain more comprehensive material information.

1.2.2   Millimeter Wave and Terahertz Spectroscopy Technologies

Millimeter wave and terahertz spectroscopy techniques study the frequency range between microwaves and infrared light in the electromagnetic spectrum, with wavelengths between 1 and 0.1 mm. Electromagnetic radiation in this band has the ability to penetrate non-conductive materials such as clothing, plastics, paper., but does not penetrate moisture and metal, so it has potential application value in fields such as security inspection, non-destructive testing and biomedical imaging [13].

The spectral characteristics of the terahertz band make it a powerful tool for analyzing complex molecular structures, such as the three-dimensional structure of proteins and the crystal structure of drug molecules. Because terahertz wavelengths can resonate with intermolecular interactions, terahertz spectroscopy technology can detect small changes at the molecular level, which is crucial for research in early disease diagnosis, materials science, and chemical reaction monitoring [14].

However, the development of millimeter-wave and terahertz spectroscopy technologies faces significant technical challenges, especially in the manufacturing of terahertz wave sources and detectors. Current terahertz wave sources have limited power and often require low-temperature operating environments, which limits their practicality. Similarly, effective detection technology in the terahertz band also requires further research and development to improve the detection sensitivity and signal-to-noise ratio of signals [15].

In addition, the atmospheric absorption characteristics of the terahertz band mean that the effective distance of terahertz transmission and communication is limited in outdoor environments. Therefore, developing adaptable terahertz systems for different application environments is another focus of researchers [16].

With the continuous advancement of materials and devices related to terahertz technology, the application prospects of terahertz spectroscopy technology in fields such as medical diagnosis, drug testing, non-destructive testing, and wireless communications are highly anticipated.

1.2.3   Application Examples of Spectral Extension

The extension of the spectrum has opened up many new application areas, covering everything from daily life to high-tech industries. Here are some notable application examples:

Security detection: The application of millimeter wave and terahertz imaging technologies in airport security inspections has become increasingly common. These imaging systems can see through passengers' clothing and detect hidden items without invading privacy, providing an effective means of security inspection [17].

Biomedical imaging: Terahertz spectroscopy technology shows great potential in early diagnosis of cancer and imaging of biological tissues. Because cancer cells and normal cells have different absorption characteristics in the terahertz band, terahertz imaging can help doctors identify tumor tissue without pathological sections [18].

Material testing: In the field of non-destructive testing, non-visible light can be used to detect defects and corrosion conditions inside materials, which is especially important for the protection of composite materials and cultural heritage [19].

Astronomy research: Astronomers use submillimeter telescopes to detect cold stars and molecular clouds in the universe. The spectral information in these bands helps to study the chemical composition of interstellar matter and the formation process of the early structure of the universe.

Communication technology: As the demand for wireless data transmission continues to grow, the terahertz frequency band is expected to become one of the key technologies for the next generation of ultra-high-speed communication networks. Although it is affected by atmospheric absorption, its high data transmission rate still has great development potential [20].

Food testing: The application of infrared spectroscopy technology in food safety monitoring is gradually increasing, such as detecting the moisture content, degree of spoilage and residues of harmful substances in food [21].

1.2.4   Future Directions and Perspectives

In the field of spectrum continuation, future research directions and prospects show a diversified trend. Scientists are working to overcome the limitations of existing technologies while exploring new areas of application. The following are some research directions and potential development trends worthy of attention:

Technology integration and miniaturization: With the continuous advancement of nanotechnology and microelectromechanical systems (MEMS), the integration and miniaturization of millimeter wave and terahertz technologies have increasingly become the focus of technology development. In the future, we can expect to see more portable, more integrated spectroscopy systems appear on the market. The reduction in size and weight of these systems will greatly facilitate their use in everyday consumer electronics, such as smartphones, wearable devices, and home security systems.

Signal processing and data analysis: In an era driven by big data, advances in signal processing and data analysis will be crucial to the development of spectroscopy technology. The integration of machine learning and artificial intelligence will improve the automation of spectral data processing and the accuracy of analysis, thereby accelerating the extraction of useful information from massive spectral data. This advancement will have important applications in fields such as pharmaceuticals, medical imaging, and environmental monitoring.

Development of new materials and devices: Research on new materials and devices is a key factor in promoting the development of spectrum extension technology. Researchers are exploring new semiconductor materials suitable for the terahertz band, superconducting materials, and optical crystals that can operate in a wider band. The emergence of these new materials and devices is expected to greatly improve the performance of spectroscopic systems, including improving sensitivity, resolution, and spectral range [22].

Integration of spectral imaging technologies: The combination of multispectral, hyperspectral and terahertz imaging will bring new dimensions to imaging technology. This fusion of technologies can provide richer information than a single spectral technology, with higher contrast and resolution capabilities. In fields such as medical development, precision agriculture, and artwork identification, multi-modal imaging systems are gradually becoming a hot research topic.

Development of communication technology: As 5G communication technology gradually matures, millimeter wave has become one of the key technologies to improve the speed and bandwidth of wireless communication. Further extension to the terahertz frequency band is expected to lay the foundation for the development of 6G and even higher-generation communication networks. Although currently facing the challenges of atmospheric absorption and path loss, terahertz communication has become a popular research direction due to its potential high-speed data transmission capabilities [20].

Sustainable development and environmental protection: Spectral continuation technology has great potential in environmental monitoring. From monitoring atmospheric pollutants and greenhouse gas emissions to assessing ecosystem health, spectroscopic technology can provide important data support. In the context of increasingly severe global climate change, spectrum extension technology will play a more important role in future environmental protection and sustainable development policy formulation.

Overall, the future prospects of spectrum extension technology are promising and it is expected to achieve revolutionary breakthroughs in many fields. With the deepening of scientific research and the strengthening of interdisciplinary cooperation, more and more innovative spectroscopic technologies will appear in the future to meet the needs of social development.

1.3 Ubiquitous Intelligent Sensing

1.3.1 Advances in Intelligent Sensing Technology

The concept of ubiquitous smart sensing encompasses a vision in which sensing technology is seamlessly integrated into every aspect of our lives, providing continuous real-time data that enhances our understanding of and interaction with our environment. Intelligent sensing technology has made significant progress, and breakthroughs in sensor miniaturization, sensitivity, selectivity, and connectivity have paved the way for smart environments [23].

Miniaturization is a key trend in sensing technology, which is leading to the development of microsensors and even nanosensors that can be widely deployed without being noticeable. These tiny sensors can be embedded into everyday objects, infrastructure, wearable devices, and even inside the human body for health monitoring. They are capable of collecting a variety of data types including physical, chemical, biological and environmental parameters.

Sensitivity and selectivity are also crucial properties, and they have been significantly improved. Sensors are not only able to detect trace amounts of substances but are increasingly able to differentiate between different stimuli with high precision. This is particularly important in applications such as pollution monitoring, where the ability to detect low concentrations of specific toxic compounds can provide early warning of environmental hazards.

Connectivity is another area where smart sensing technology is making advances. With the advent of the Internet of Things (IoT) and wireless sensor networks, sensors can now communicate with each other and with central systems, making it possible to obtain data from large networks of distributed sensors. Aggregate data. This connectivity facilitates large-scale data analysis and plays a role in applications such as smart cities, where sensors monitor traffic flow, energy usage and other critical infrastructure parameters.

Machine learning and artificial intelligence (AI) are increasingly integrated with sensing technologies to enable intelligent decision-making. By processing and analyzing sensor data, these systems can learn patterns, predict outcomes, and automate responses. For example, in precision agriculture, sensors can detect soil moisture and nutrient levels, and AI algorithms analyze the data to optimize irrigation and fertilization schedules.

Energy efficiency and collection have become important areas of focus for smart sensors. The pursuit of sustainability has led to the development of sensors with minimal power requirements that can operate on energy harvesting technologies such as solar or kinetic energy and even extract energy from the signals they measure. This advance is critical for deploying sensors in remote areas and in applications where replacing batteries is impractical.

In summary, advances in smart sensing technology facilitate the transition to a world where sensors are ubiquitous and critical to data-driven decisions. The integration of these sensors with advanced analytics and AI is transforming industry, healthcare, environmental monitoring and our daily lives, bringing the promise of smarter, more efficient and more responsive systems.

1.3.2   Overview of Applications Across Multiple Domains

The proliferation of smart sensing technologies has led to a wide range of applications across many fields, providing critical insights and making decision-making processes smarter. The applications of these technologies are diverse and impact almost every sector of society. The following is an overview of smart sensing applications in different fields:

Healthcare: In the medical field, wearable sensors monitor important vital signs such as heart rate, blood pressure, and blood sugar in real time, providing patients with continuous health data. These devices can predict and alert individuals of possible health issues, allowing healthcare to be more proactive. Hospitals are also using smart sensors to track medical assets and monitor patient conditions, improving the quality of care and operational efficiency.

Automotive Industry: Smart sensors are at the core of the development of autonomous vehicles. They are used for environmental perception, obstacle detection and navigation, significantly improving traffic safety and efficiency. In addition, on-board sensors monitor driver alertness, vehicle performance, and provide predictive maintenance reminders, adding more intelligence and adaptability to the driving experience.

Environmental monitoring: Smart sensors deployed in various ecosystems can detect changes in climate conditions, pollution levels, and track wildlife activity. They provide valuable data for climate research and environmental protection planning, and are critical in disaster response scenarios, such as providing early warning systems for events such as earthquakes and floods.

Smart cities: Smart sensing is the cornerstone of smart city infrastructure. Sensors manage and optimize a building's energy consumption, monitor air and water quality, and simplify waste management. Intelligent transportation systems reduce congestion, improve road safety, and facilitate the implementation of smart parking solutions to make urban environments more sustainable.

Agriculture: Precision agriculture relies on smart sensors to monitor soil conditions, crop health and weather patterns. By providing precise data, these sensors enable farmers to make informed decisions about irrigation, fertilization and harvesting, optimizing resource use and increasing yields while minimizing environmental impact.

Manufacturing: In the industrial sector, smart sensors are critical for predictive maintenance, able to detect equipment anomalies that could lead to failure. They also monitor manufacturing environments and production processes, ensure quality control, and improve efficiency through real-time data analysis.

Retail and consumer goods: Smart sensors in retail environments track inventory levels, foot traffic and shopping behavior, allowing retailers to optimize store layout and inventory management. In consumer products, sensors in smart home appliances provide users with usage data, maintenance reminders, and are integrated with smart home systems to enhance convenience.

Entertainment and Leisure: The entertainment industry is using smart sensors to create immersive and interactive experiences. For example, sensors track motion and gestures in virtual reality systems to provide users with a more engaging experience. In sports and fitness, sensors monitor performance and provide athletes and enthusiasts with feedback to improve their training programs.

The applications of smart sensing are widespread and constantly expanding. With the development of technology, the integration of these sensors in various fields not only improves efficiency and productivity, but also promotes the improvement of quality of life and environmental sustainability.

1.3.3 Agriculture and Environmental Monitoring

Integrating smart sensing technology into agriculture and environmental monitoring is driving innovation in these fields by providing accurate, real-time data, promoting efficiency and sustainability. These advances have led to better resource management, increased crop yields, reduced environmental impacts, and more informed environmental policy decisions.

Agriculture: Smart agriculture is one of the most prominent areas for smart sensing applications. A variety of sensors are used to monitor soil moisture, pH, temperature, humidity and nutrient content. Drones equipped with multispectral cameras can capture aerial images to help assess plant health, detect pests and diseases, and optimize the use of water and fertilizer. Smart sensors enable precision agriculture, reducing waste and environmental damage by allowing farmers to precisely apply the required water, fertilizers, and pesticides [24].

The livestock industry also benefits from smart sensing technology, with sensors monitoring the health of animals. These sensors track location, activity levels and biological data to provide insights into livestock health and alert farmers to issues that require attention. This has led to improvements in animal husbandry practices and overall farm productivity.

Environmental monitoring: Smart sensors play a key role in environmental protection and climate change research. Networks of sensors placed in different habitats can monitor air and water quality, greenhouse gas emissions and biodiversity. These ongoing streams of data help scientists understand ecological changes and develop strategies to protect endangered species and habitats.

Remote sensing technologies, such as satellite imagery and aerial drones, allow widespread monitoring of the effects of deforestation, land use change and natural disasters. These technologies enable rapid assessment of affected areas and are critical for timely and effective disaster response and management.

In addition, smart sensors are also indispensable for urban environmental monitoring. Smart city initiatives rely on these sensors to measure urban air quality, noise pollution, and water supply conditions [25]. The data collected by these sensors supports the development of policies to reduce pollution, manage traffic congestion and promote sustainable urban development.

In conclusion, the application of smart sensing in agriculture and environmental monitoring provides comprehensive data that can bring about changes in the way we manage natural resources and respond to environmental challenges. With these technologies developing, they offer the potential to achieve greater precision and insight, driving further improvements in sustainability and environmental management.

1.3.4 Medical Health and Biometrics

Advances in smart sensing in healthcare and biometrics are driving a new era of personalized medicine and health monitoring. These technologies are shaping the way we understand human physiology, diagnose disease, and manage patient care. The integration of smart sensors in the medical field not only improves patient treatment effects, but also simplifies medical procedures, reduces costs, and improves the overall quality of medical services.

Healthcare: Smart sensing technology is critical to the development of advanced diagnostic tools and treatment devices. Wearable sensors are increasingly used to monitor chronic conditions such as diabetes, heart disease and respiratory diseases. They provide real-time data streams that can be analyzed to detect signs of worsening or the onset of an acute episode, allowing for timely intervention.

Implantable sensors provide a deeper understanding of patient health by monitoring internal physiological parameters. These sensors can track everything from heart rhythm to blood sugar levels, providing a complete picture of a patient's health without the need for invasive surgery. Additionally, sensor-equipped smart pills are being explored that have the potential to transmit data from within the gastrointestinal tract, providing valuable insights into digestive health and drug absorption.

Biometrics: Smart sensing is also reshaping the field of biometrics, which exploits physiological and behavioral characteristics for identification and security purposes. Fingerprint sensors, facial recognition, iris scanning, and voice recognition are now widely used in consumer electronics for security authentication [26]. In addition to security, biometric sensors in medical environments can help confirm patient identity, reduce the risk of medical errors, and improve patient safety.

Telemedicine and remote patient monitoring are becoming increasingly important in providing care to patients in remote locations. Smart sensors play a key role here, collecting health data that can be transmitted to medical providers in real time, allowing them to make informed decisions remotely. This approach would be particularly beneficial in elderly care, where continuous monitoring could alert caregivers before health problems escalate.

Additionally, smart sensing is critical to advancing robotic surgery and assisted living technologies. In robotic surgery, sensors provide feedback to the surgeon on force and positioning, improving the accuracy of surgical procedures [27].

Assisted living technology, such as fall detection systems and automated medication dispensers, rely on sensors to ensure the safety and independence of people with mobility or cognitive impairments.

Research and clinical trials: In clinical research, smart sensors are revolutionizing the way clinical trials are conducted, enabling continuous monitoring of participants and real-time data collection. This results in a more accurate and complete data set that can improve the accuracy of the evaluation of new medical treatments and therapies.

The future of healthcare and biometrics is closely tied to smart sensing technology. As these technologies continue to evolve, they will support more complex applications such as real-time monitoring of multiple biomarkers, integration with artificial intelligence for predictive analytics, and the development of fully personalized treatment options. These technologies have enormous potential to improve patient care and medical outcomes and mark an important step in the development of modern medicine.

In summary, smart sensing technologies are transforming healthcare and biometrics, providing unprecedented opportunities to improve patient care, advance medical research, and enhance the overall healthcare experience. The capabilities of these sensors will continue to grow, supporting new and innovative applications that will further revolutionize healthcare in the coming years.

1.3.5   Industrial and Security Inspection

Smart sensing technologies have become a key component in the industrial and security sectors, where they improve safety, efficiency and reliability. In the industrial field, the application of smart sensors is often regarded as part of the Industrial Internet of Things (IIoT), which is driving the fourth industrial revolution, the so-called Industry 4.0 [28]. In the security world, smart sensing helps protect assets, spaces, and individuals by providing advanced surveillance and detection capabilities.

Industrial Detection: Smart sensors are widely used in industrial environments to monitor machinery and infrastructure. These sensors collect data on temperature, vibration, pressure, and other critical parameters that may indicate the condition of the equipment [29]. By implementing predictive maintenance strategies using these sensors, industry can prevent unplanned downtime, extend equipment life, and optimize maintenance operations. For example, vibration sensors can detect irregularities in machinery, which may indicate the need for repairs before failure occurs.

Additionally, smart sensors are critical to ensuring the quality and integrity of industrial processes. They are used on production lines to check products for defects or deviations from standards. Advanced imaging sensors and machine vision technology identify the tiniest imperfections, resulting in higher quality products and lower production scrap rates.

Safety in hazardous environments such as chemical plants or refineries is another area where smart sensors are indispensable. Gas leak detectors, flame detectors and other types of sensors can provide early warning of potential hazards, allowing rapid evacuation and response measures to prevent an accident or limit its impact [30].

Security detection: In the security field, smart sensing technology is used in a range of applications from access control to perimeter defense. For example, biometric sensors are now widely used in high-security areas to ensure that only authorized individuals have access. Surveillance cameras equipped with intelligent image analysis can detect unusual activity or behavior and trigger alarms for security personnel.

Advanced sensors also play an important role in detecting contraband or hazardous materials. Scanning systems using X-ray, terahertz and other spectroscopic methods can non-invasively inspect luggage, cargo and vehicles for threats, enhancing security at airports, border crossings and other critical infrastructure.

The integration of smart sensing with other technologies such as artificial intelligence (AI) and big data analytics further amplifies its potential in industrial and security applications. AI algorithms can analyze large amounts of sensor data, identify patterns, predict outcomes, and automate the decision-making process. This integration allows more sophisticated monitoring systems to adapt and respond to changing conditions in real time.

In summary, smart sensing is reshaping the industrial and security inspection landscape, providing enhanced capabilities for predictive maintenance, quality control, security management, and threat detection. As these technologies continue to advance, they will play an even greater role in ensuring the smooth functioning of industry and the importance of protecting people and assets from a wide range of risks.

1.3.6 Intelligent Cities and Traffic Management

Smart cities rely on intelligent sensing technology to enhance the quality of urban life and improve urban operations. These technologies involve building a network of sensors and Internet of Things (IoT) devices that are interconnected through data analysis systems to optimize urban services such as traffic management, public transportation, energy use, and environmental monitoring [31, 32].

Traffic Management: Smart sensors play a key role in traffic management in smart cities. Sensors are embedded in the infrastructure and collect data on vehicle movements, congestion levels and traffic patterns. This data is analyzed in real time to adjust traffic lights, reduce congestion and improve commute times [31]. Intelligent transportation systems can also provide drivers with warnings of road incidents and alternative routes, improving road safety and efficiency.

Public transportation: In public transportation, smart sensors facilitate tracking of vehicles, managing timetables and assessing passenger flow. This improves service reliability and planning capabilities. Contactless payment systems powered by smart sensing technology speed up transaction times and improve user experience [31].

Energy consumption: Smart cities use smart sensors to manage and conserve energy more efficiently. Smart grids equipped with sensors can detect and respond to changes in energy demand, improve energy distribution and integrate renewable energy more efficiently [32].

Environmental monitoring: Smart sensors in smart cities are used to track environmental factors such as air and water quality, noise pollution, and more. This real-time data helps inform policies aimed at reducing pollution and mitigating the effects of climate change.

Urban planning: Data generated by smart sensing technology provides valuable insights into urban development. Planners can use this data to make informed decisions about infrastructure investment and resource management. Data can also help improve public safety by identifying areas that may need improved lighting or signage.

In summary, smart sensing is the core of the smart city paradigm, providing solutions for modernizing transportation systems, improving public services, and promoting sustainable development. As cities continue to expand, the role of these technologies becomes even more important. Smart cities equipped with smart sensing have the potential to become more efficient, sustainable and provide a higher quality of urban life, thereby better addressing the challenges posed by increasing urbanization.

1.3.7 Challenges and Future Trends

Smart sensing technology brings many benefits across multiple areas, but also poses some challenges that must be addressed to fully realize its potential. Furthermore, the field of smart sensing continues to evolve, with new trends emerging that will shape the future of this dynamic field.

Challenge:

Data Privacy and Security: As smart sensors collect large amounts of data, privacy and security issues become even more important. Securing sensitive data from unauthorized access and cyberattacks is critical.

Standardization: As more devices and sensors are deployed, standardization is needed to ensure compatibility and interoperability between different systems and platforms.

Energy Consumption: The energy requirements to power a large sensor network can be considerable. Developing sensors with low power consumption or energy harvesting is critical for sustainable and scalable deployment.

Data management: The vast amounts of data generated by smart sensors can overwhelm existing data processing and storage capabilities. Efficient algorithms and infrastructure are needed to manage this data and extract meaningful insights from it.

Integration Complexity: Integrating smart sensors with existing systems and ensuring they work harmoniously with legacy devices is a complex task that requires careful planning and execution.

Future trends:

Edge Computing: To address data management challenges, there is a growing trend toward edge computing, where data processing occurs closer to the source of data collection, reducing the need to transfer large amounts of data to central servers.

Artificial Intelligence and Machine Learning: The trend of integrating Artificial Intelligence and Machine Learning with smart sensing is expected to increase, enabling more advanced analytics, predictive maintenance, and autonomous decision-making.

Advanced Materials: Developments in new materials and nanotechnology may lead to the creation of more sensitive, more accurate and more durable sensors, opening up new application areas and improving existing ones.

5G Connectivity: The rollout of 5G networks will significantly enhance the capabilities of smart sensors by providing higher speeds, reduced latency and the ability to connect a large number of devices simultaneously.

Sensory Internet: In addition to the Internet of Things, the concept of the Sensory Internet is emerging, which involves the use of sensors to digitally simulate human senses, potentially leading to more immersive and interactive experiences.

Smart sensing technology is expected to continue to have a transformative impact in numerous fields. By solving current challenges and embracing upcoming trends, the field can move forward toward a future where smart sensing is more integrated into our daily lives, driving innovation and improving quality of life around the world.

1.4   The Practice of Safe Rural Areas

The role of intelligent security in the harmonious development of the nation and the stability of social safety is becoming increasingly prominent. The new generation of intelligent security systems is characterized by a broad scope of application, a large user base, high video resolution, and complex transmission environments. With China's digital rural transformation, the demand for video monitoring has surged significantly, with the number of access channels for safe rural area projects exceeding 25 million in 2021.

The deployment of monitoring cameras in rural areas is on the rise, and the development and intelligence of end-side sensors are becoming more and more important. To adapt to the differentiated scenarios in rural areas, preserve more key scene information, and achieve optimal solutions tailored to rural settings, an important example is the deployment of active vision cameras at the end side. Unlike passive video recording cameras, active vision cameras are capable of active perception and computation, adapting to changes in the scene and processing key information locally within the camera.

For the safe rural area monitoring scenarios (home monitoring, production monitoring, public area monitoring), image recognition and rapid alert algorithms are becoming increasingly important. This includes applications such as home protection, elderly care, theft and fire prevention, and field and forest monitoring.

Here is the introduction of a security monitoring system designed for rural use:

The Netgear Video Doorbell is a product from Arlo Technologies (Fig. 1.1), a home security camera company under the American networking equipment manufacturer Netgear. Arlo is a highly successful product line.

The "Safe Rural Area" video monitoring platform, relying on China Telecom's Tianyi Cloud Monitoring Service (Fig. 1.2), has become an effective means of gaining first-hand material on rural sites, capturing a vast amount of real-time and dynamic image information. It allows around-the-clock monitoring of vehicles' entries and exits and the movement of people without the need to send personnel on rotational guard duties, enhancing the sense of security for the public and the level of comprehensive governance and safety management in rural areas. During the epidemic, the "Safe Rural Area" remote monitoring has acted as a veritable "thousand-mile eye," establishing a "cloud guardianship" model for epidemic prevention and control

Fig. 1.1 Netgear video doorbell from Arlo Technologies, a home security camera

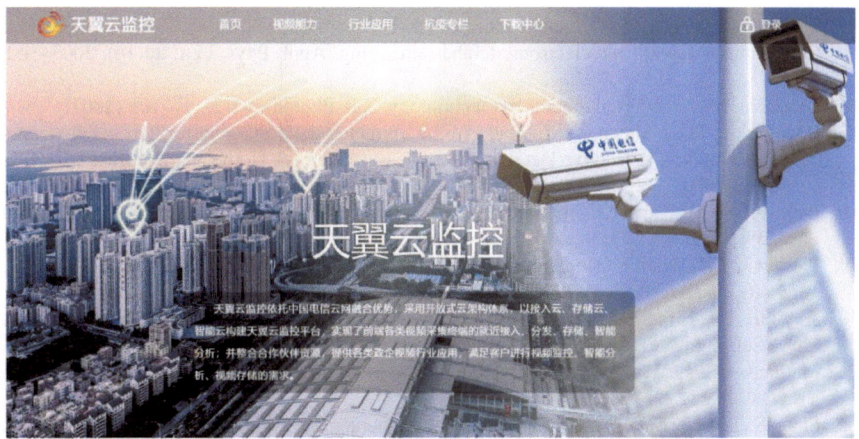

Fig. 1.2 China Telecom's Tianyi Cloud Monitoring Service

in rural areas. At the same time, the "Safe Rural Area" monitoring system innovatively realizes intelligent analysis and assists in actual operations, providing technical support for precise flood prevention and risk resistance, scientific decision-making and command in towns and villages.

References

1. Liaghat, S., Balasundram, S.K.: A review: the role of remote sensing in precision agriculture. Am. J. Agric. Biol. Sci. **5**(1), 50–55 (2010)
2. Gitelson, A.A., Kaufman, Y.J., Merzlyak, M.N.: Use of a green channel in remote sensing of global vegetation from EOS-MODIS. Remote Sens. Environ. **58**(3), 289–298 (1996)
3. Berns Roy, S., Billmeyer, S.: Principles of Color Technology (2000)
4. Curran, P.J.: Multispectral remote sensing for the estimation of green leaf area index. Philos. Trans. R. Soc. Lond. Ser. B Biol. Sci. **324**(1223), 247–262 (1989)
5. Vane, G., Goetz, A.F.H.: Terrestrial imaging spectrometry: current status, future trends. Remote Sens. Environ. **44**, 117–126 (1993)
6. Bioucas-Dias, J.M., et al.: Hyperspectral remote sensing data analysis and future challenges. IEEE Geosci. Remote Sens. Mag. **1**(2), 6–36 (2013)
7. Eismann, M.T.: Hyperspectral Remote Sensing (2012)
8. Camps-Valls, G., Bruzzone, L.: Kernel-based methods for hyperspectral image classification. IEEE Trans. Geosci. Remote Sens. **43**(6), 1351–1362 (2005)
9. Lennon, R.: Remote Sensing Digital Image Analysis: An Introduction. ESA/Esrin (2002)
10. Socrates, G.: Infrared and Raman Characteristic Group Frequencies: Tables and Charts. Wiley (2004)
11. Griffiths, P.R.: Fourier transform infrared spectrometry. Science **222**(4621), 297–302 (1983)
12. Griffiths, P.R. (ed.): Handbook of Vibrational Spectroscopy, vol. 4. J. M. Chalmers Wiley (2002)
13. Mittleman, D. (ed.): Sensing with Terahertz Radiation, vol. 85. Springer (2013)
14. Siegel, P.H.: Terahertz technology in biology and medicine. IEEE Trans. Microw. Theory Tech. **52**(10), 2438–2447 (2004)

15. Siebert, K.J., Quast, H., Leonhardt, R., Löffler, T., Thomson, M., Bauer, T., Czasch, S., et al.: Continuous-wave all-optoelectronic terahertz imaging. Appl. Phys. Lett. **80**(16), 3003–3005 (2002)
16. Tonouchi, M.: Cutting-edge terahertz technology. Nat. Photonics **1**(2), 97–105 (2007)
17. DMcMakin, D.L., Sheen, D.M., Hall, T.E.: Millimeter-wave imaging for concealed weapon detection. In: Nondestructive Detection and Measurement for Homeland Security, vol. 5048, pp. 52–62. SPIE (2003)
18. Pickwell, E., Wallace, V.P.: Biomedical applications of terahertz technology. J. Phys. D: Appl. Phys. **39**(17), R301 (2006)
19. Piesiewicz, R., Kleine-Ostmann, T., Krumbholz, N., Mittleman, D., Koch, M., Kürner, T.: Terahertz characterisation of building materials. Electron. Lett. **41**(18), 1002–1004 (2005)
20. Akyildiz, I.F., Jornet, J.M., Han, C.: Terahertz band: next frontier for wireless communications. Phys. Commun. **12**, 16–32 (2014)
21. Mathanker, S.K., Weckler, P.R., Wang, N.: Terahertz (THz) applications in food and agriculture: a review. Trans. ASABE **56**(3), 1213–1226 (2013)
22. Shen, S., Liu, X., Shen, Y., Qu, J., Pickwell-MacPherson, E., Wei, X., Sun, Y.: Recent advances in the development of materials for terahertz metamaterial sensing. Adv. Opt. Mater. **10**(1), 2101008 (2022)
23. Zanella, A., Bui, N., Castellani, A., Vangelista, L., Zorzi, M.: Internet of things for smart cities. IEEE Internet Things J. **1**(1), 22–32 (2014)
24. Güven, B., Baz, İ., Kocaoğlu, B., Toprak, E., Erol Barkana, D., Soğutmaz Özdemir, B.: Smart farming technologies for sustainable agriculture: from food to energy. In: A Sustainable Green Future: Perspectives on Energy, Economy, Industry, Cities and Environment, pp. 481–506. Springer International Publishing, Cham (2023)
25. Ramírez-Moreno, M.A., Keshtkar, S., Padilla-Reyes, D.A., Ramos-López, E., García-Martínez, M., Hernández-Luna, M.C., Lozoya-Santos, J.D.J., et al.: Sensors for sustainable smart cities: a review. Appl. Sci. **11**(17), 8198 (2021)
26. Guo, Y., Liu, X., Peng, S., Jiang, X., Xu, K., Chen, C., Chen, W., et al.: A review of wearable and unobtrusive sensing technologies for chronic disease management. Comput. Biol. Med. **129**, 104163 (2021)
27. Bhattacharyya, D., Ranjan, R., Alisherov, F., Choi, M.: Biometric authentication: a review. Int. J. u- e-Serv. Sci. Technol. **2**(3), 13–28 (2009)
28. Xu, L.D., Xu, E.L., Li, L.: Industry 4.0: state of the art and future trends. Int. J. Prod. Res. **56**(8), 2941–2962 (2018)
29. Gubbi, J., Buyya, R., Marusic, S., Palaniswami, M.: Internet of Things (IoT): a vision, architectural elements, and future directions. Futur. Gener. Comput. Syst. **29**(7), 1645–1660 (2013)
30. Sadeghi, A.R., Wachsmann, C., Waidner, M.: Security and privacy challenges in industrial internet of things. In: Proceedings of the 52nd Annual Design Automation Conference, pp. 1–6 (2015)
31. Hancke, G.P., Hancke, G.P., Jr.: The role of advanced sensing in smart cities. Sensors **13**(1), 393–425 (2013)
32. Batty, M., Axhausen, K.W., Giannotti, F., Pozdnoukhov, A., Bazzani, A., Wachowicz, M., Portugali, Y., et al.: Smart cities of the future. Eur. Phys. J. Spec. Top. **214**, 481–518 (2012)

Chapter 2
Active Vision Camera System

Abstract In this chapter, we propose an intelligent camera system based on active vision sensing technology. It is able to capture visible and near-infrared signals using ambient light, and combines algorithms and embedded hardware at the front end, thus solving the challenges of traditional cameras with limited field of view and monitoring targets, and color distortion. We describe in detail the adaptive exposure control scheme, the image processor adapted to RGB-IR, and the flexible task scheduling mechanism in the active vision camera system. In experiments and practical applications, the intelligent sensor camera system significantly improves the quality of image acquisition in foggy and low-light environments at night, especially improving the effect of face detection.

Keywords Active vision sensing · Intelligent camera · Adaptive exposure control · RGB-IR image processor · Dynamic scheduling

2.1 Design Concept of Active Visual Sensing

Active vision intelligent sensing technology has expanded single-station observation from the conventional range of several hundred meters to tens of kilometers, solving the challenges faced by traditional cameras in meeting the requirements for wide-area and multi-target monitoring [1] due to limited sensing resolution and spectral quality.

Conventional camera technology is affected by factors such as motion, lighting, and noise, which greatly limit the accuracy of the data captured by intelligent cameras. The RGB-IR hybrid pixel array technology has been proposed to capture signals in the visible light and near-infrared bands using ambient light. This innovation has realized active vision intelligent cameras that perfectly solve the issue of color distortion in over-bright pixels while capturing richer information. The specialized intelligent front-end, which integrates smart capture, automatic image filtering, live detection, automatic exposure adjustment, and wide dynamic range functions, is designed holistically by combining algorithm research with embedded hardware system design. It balances multiple factors such as the capture environment, image quality, computing power, and power consumption. Experimental results show that, compared to the most advanced intelligent cameras, the end-side active intelligent sensing cameras that we propose significantly improve image quality in nighttime conditions with illumination, nighttime without illumination, and in foggy weather. Taking the face detection task as an example, there are noticeable improvements in face detection speed, detection rate, and recognition accuracy.

2.2   Adaptive Exposure Control to the Environment

Currently, due to the limited dynamic range of image sensors, it is challenging to capture details in both bright and dark areas within a scene with strong contrast between light and shadow during imaging. This has led to the development of Wide Dynamic Range (WDR) technology [2, 3], which allows for cap Existing WDR technologies mainly include two types: one is the image fusion with two exposures implemented within the sensor, and the other is the image fusion with two exposures processed on the processor side, resulting in an image with a wider dynamic range.

The advantage of both methods is that they aim to achieve superior WDR effects in a single image. They are especially suitable for monitoring applications where only one camera is used, as well as for back-end image processing and analysis applications, which can be realized by purchasing sensors with WDR capabilities or cameras with WDR fusion algorithms.

However, when these methods are used for image analysis, only the fused image can be obtained. The image fusion under two exposures often leads to an uneven transition of brightness information between overexposed high-light areas and underexposed low-light areas, which can cause inaccuracies in object detection at their boundaries. Moreover, for cameras used solely for object detection, the choices are limited to selecting image sensors with WDR functions or purchasing cameras with WDR algorithms. Such cameras, due to the image fusion algorithm and the encoding/ decoding transmission, can cause significant delays in image computation (at least 4 frames), affecting the overall speed of object detection and recognition.

Therefore, it is vital to address the limitations, slow speed, and low accuracy of existing WDR technologies when used for object detection [4]. To overcome these drawbacks and achieve fast and accurate object detection in images, this design is based on traditional non-WDR image sensors and achieves WDR effects through multiple exposures within an embedded system. It involves receiving an image to be detected from the image sensor, calculating the average brightness of the entire image and the average brightness of specific zones, and determining the exposure parameters for one or more subsequent frames to be detected. Based on the exposure parameters for the subsequent frame(s), the system then captures one or more frames to be detected. In each instance, after capturing the target image, the system performs object detection steps to obtain the detection results, which include subsequent frame(s) to be detected, thereby achieving WDR effects through these steps.

Turing more comprehensive images where details in both high brightness areas (such as under strong light sources) and low brightness areas (such as shadows or backlighting) can coexist, facilitating object detection across the entire scene.

2.2.1 Overview of Multiple Exposure Wide Dynamic Range Technology

The implementation process for multiple exposure wide dynamic range technology is as follows (Fig. 2.1):

1. Start with the nth frame as the first received frame. Calculate the average brightness of the entire image from this frame and determine the optimal exposure parameters for an average brightness level, which are then sent to the image sensor.
2. With the exposure parameters from step 1, obtain the (n + 1)th frame. This frame is used to calculate both the average brightness of the entire image and the average brightness of individual zones. For instance, with a commonly used resolution

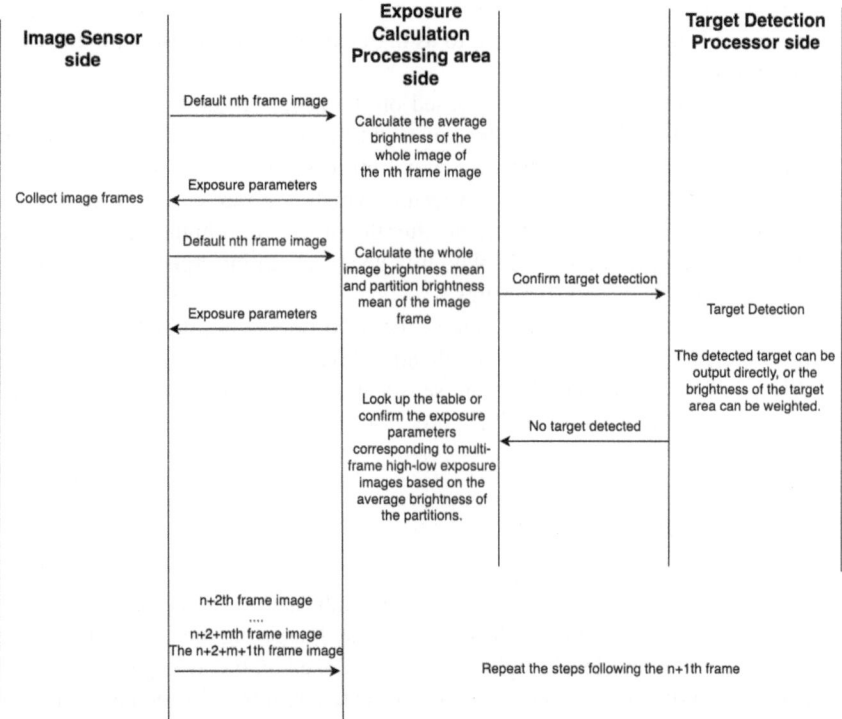

Fig. 2.1 Wide dynamic range implementation for target detection

of 1920×1080, divide the image into 16×9 zones, with each zone consisting of 120×120 pixels, and calculate the averaging brightness for each of these zones. The purpose of this zonal brightness calculation is to identify any overexposed bright areas or underexposed dark areas in the image while knowing their exact locations.

3. If the brightness of all zones in the $(n + 1)$th frame does not significantly differ from the whole image, use the whole image's average brightness to calculate the exposure parameters. If there are no overexposed bright areas or underexposed dark areas in the image, then proceed with object detection for the entire image. If an object is detected, further calculate the average brightness within the target zone, and subsequent exposures can be adjusted with weighted brightness for that specific area.

4. If the $(n + 1)$th frame has zones with brightness exceeding the high or low thresholds, calculate the exposure parameters for subsequent frames. Based on the zonal brightness values, you can decide whether to increase or decrease exposure. If the threshold is exceeded with a stepped change, use a lookup table to adjust the exposure by two or three steps to capture more image details in various brightness levels. If there are both overexposed bright areas and underexposed dark areas, alternatingly collect images with increased and decreased exposures based on the average brightness of the whole image—increased exposure for dark areas and decreased exposure for bright areas. Subsequent exposures with increased brightness are referred to as high exposure, and those with decreased brightness as low exposure.

5. Following the $(n + 1)$th frame, based on the zonal brightness calculations, m frames might be collected, where m could be as low as one frame (when the brightness in all zones is balanced) and up to six frames (three steps of high exposure + three steps of low exposure, which is a rare case). After this, the calculation can loop from the frame after the $(n + 1)$th. If high or low exposures are present and needed for display, it is necessary to intersperse some average whole image brightness exposures.

6. Starting with the nth frame is due to the need for system initialization, which precedes image reception. The application layer starts after the entire machine driver is loaded, during which the sensor has already output n frames, but the application layer has not received, processed, or analyzed them.

2.2.2   Detail Explanation

As outlined in Step 1 of Sect. 2.2, the first image (nth frame) is subjected to automatic exposure. An average brightness level is calculated targeting the highest brightness value of 255 and a mid-level brightness of 128. Based on the average brightness of the entire nth frame image, the exposure parameters required for the next frame are calculated. These parameters are used to obtain the $(n + 1)$th target image with a well-balanced overall brightness level.

After each target image is captured, object detection steps are performed, which specifically involves: calculating the difference in brightness between the average brightness of the entire image and that of each divided zone. The difference between the overall image brightness average and the zone brightness averages is used to distinguish between the first and second target detection modes. If the difference is small, it is classified under the first target detection mode; if the difference is large, it is classified under the second target detection mode.

In cases where the target image has a suitably exposed auto-exposure, calculate the average brightness for both the entire image and divided zones of the initially received image. If the difference between them is small, classify it as the first target detection mode and conduct a full-area target detection across the entire image. If the difference is substantial, classify it as the second target detection mode, and perform target detection in the zones with balanced brightness. In addition, perform multi-stage exposure in the excessively bright or dark areas to capture images with normal brightness levels in these zones, thereby completing target detection across the entire image. Each complete full-area target detection is considered a cycle. Within a cycle, it may be necessary to capture only one image, corresponding to the first target detection mode, or multiple images, corresponding to the second target detection mode.

2.2.2.1 First Target Detection Mode

The steps for target detection using the first target detection mode are as follows (Fig. 2.2):

- Determine the target area in the target image; calculate the regional brightness mean of the target area; determine the next exposure parameters of the target area based on the regional brightness mean of the target area; perform target detection based on the image obtained from the next exposure parameters of the target area, and obtain the target detection result. In this mode, the difference between the average brightness of the whole image and the average brightness of the partitions of the target image is small. Within the set threshold, the target in the entire image can be detected and identified without further adjustment of the exposure.
- Although target detection can be directly conducted with the above target image and actual targets can be detected to yield results, the brightness in the region where the target is located may not be optimal and may require further adjustment. This means that after detecting the target area, weighted adjustments can be applied to further refine the exposure, optimizing the target area that has been detected.
- In more detail, delineate the area within the target image where the target is located, perform a calculation of the average brightness for that area, and obtain the average brightness value. Then, based on the average brightness of the target area, apply a weighted adjustment to the target's brightness. This determines the exposure parameters for the image to be output, after which the image sensor captures the

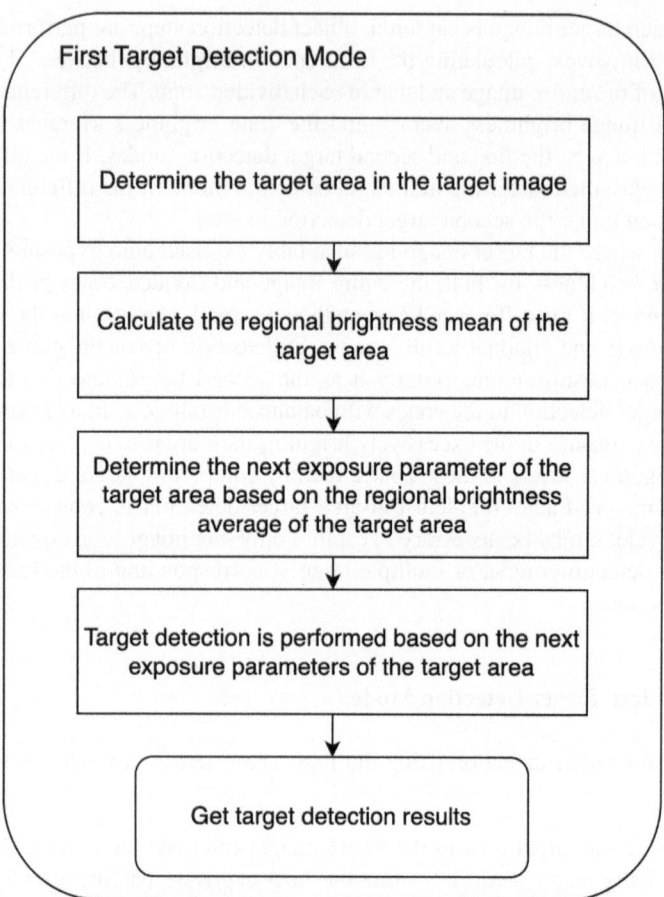

Fig. 2.2 First target detection mode

image to be output according to the determined exposure parameters, resulting in the target detection outcomes.

2.2.2.2   Second Target Detection Mode

In this mode, if the image has only a few regions where the difference between the overall image brightness mean and the sectional brightness mean exceeds the threshold, and the differences are all positive (overexposed) or all negative (underexposed), then the first mode of multi-step high exposure or low exposure is directly selected. If the differences between the overall image brightness mean and the sectional brightness mean within the image include both positive and negative values that exceed the threshold, then it is necessary to select a combined mode of multi-step high exposure and low exposure.

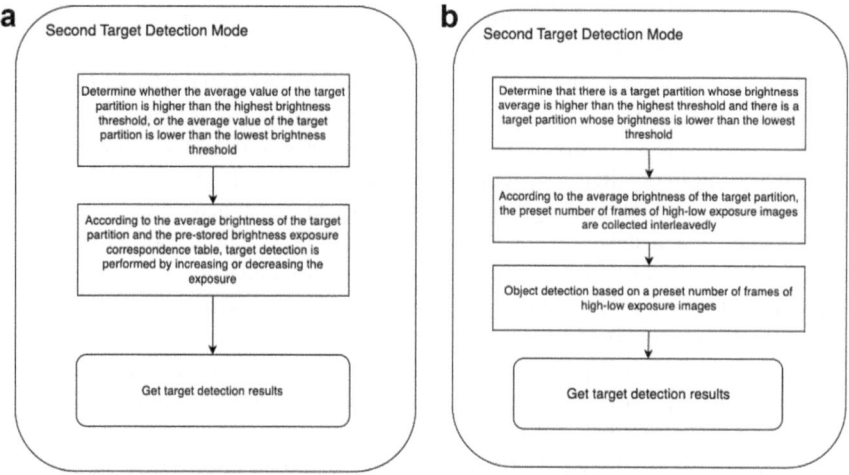

Fig. 2.3 Second target detection single (**a**) and mixed (**b**) modes

Multi-step High Exposure or Low Exposure Single Mode

As shown in Fig. 2.3, the second target detection mode is used for target detection, which includes: determining if there is a target section with an average brightness above the highest brightness threshold, or a target section with an average brightness below the lowest brightness threshold. Based on the average brightness of the target section and the pre-stored brightness-exposure correspondence table, target detection is carried out by increasing or decreasing the exposure to obtain the target detection result.

It is easy to understand that, in this specific embodiment, among all the sections divided for the target image, only the situations of high-light overexposure or low-light underexposure exist. The sections with high-light overexposure and those with low-light underexposure are marked as target sections, and the average brightness corresponding to these target sections is referred to as the target section brightness average.

Specifically, when at least one section has a target section brightness average higher than the highest brightness threshold, it means that the section with the target section brightness average is a high-light overexposure area. At this time, the target section brightness average exceeds the highest brightness threshold and shows a stepped change. The corresponding step can be found in the pre-stored brightness-exposure correspondence table, determining to decrease the exposure based on the target image and to set the corresponding exposure parameters. The image sensor captures the next frame of the image according to the determined exposure parameters and performs target detection to obtain the target detection result.

When at least one section has a target section brightness average below the lowest brightness threshold, it means that the section with the target section brightness average is a low-light underexposure area. At this time, the target section brightness average is below the lowest brightness threshold and also shows a stepped change. The corresponding step can be found in the pre-stored brightness-exposure correspondence table, determining to increase the exposure based on the target image and to set the corresponding exposure parameters. The image sensor captures the next frame of the image according to the determined exposure parameters and performs target detection to obtain the target detection result.

The highest and lowest brightness thresholds can be set according to the actual environment. For instance, for 8-bit data in RGB888 format, the brightness value range is 0–255, where the highest brightness threshold can be set to 160, and the lowest brightness threshold can be set to 30, without specific limitation here.

The brightness-exposure correspondence table stores the relationship between brightness and exposure parameters, including multiple brightness steps and their corresponding exposure parameters. Once the target section brightness average is determined, it is possible to know how many steps the target section brightness average has changed based on the overall image brightness average through the brightness-exposure correspondence table, thus determining the number of steps for low exposure or high exposure to apply on the second frame of the image (e.g., one step high/low exposure, two steps high/low exposure, or three steps high/low exposure).

The specific exposure methods can be as follows:

1. One-step high exposure + average exposure;
2. Two-step high exposure + average exposure;
3. Three-step high exposure + average exposure;
4. One-step low exposure + average exposure;
5. Two-step low exposure + average exposure;
6. Three-step low exposure + average exposure.

It should be noted that the average exposure after each step of high or low exposure is considered to be necessary for displaying the collected image frames. Before displaying the image, the image frame will undergo an overall image brightness average exposure treatment.

For a good user experience on the display end, high-exposure or low-exposure images that have not undergone overall image brightness average exposure treatment are generally not displayed.

The number and size of the brightness steps in the brightness-exposure correspondence table can be set according to the actual situation and are not specifically limited here.

For example, if the overall image brightness average of the target image is moderate at 130 and the target section brightness average is 180, assuming that each brightness step in the brightness-exposure correspondence table is 20, then the target section brightness average is on the third brightness step. At this time, three steps of low exposure can be performed based on the target image in order to obtain more detailed brightness information in the image and complete the target detection.

Multi-step High Exposure and Low Exposure Mixed Mode

As shown in Fig. 2.3, the second target detection mode is employed for target detection, which involves: determining if there is a target section with an average brightness above the highest brightness threshold, and if there is a target section with an average brightness below the lowest brightness threshold; collecting a preset number of frames of high-low exposure images alternately based on the target section brightness average; and carrying out target detection based on the preset number of frames of high-low exposure images to obtain the target detection result.

In this specific embodiment, it is understandable that among all the sections divided for the target image, situations of both high-light overexposure and low-light underexposure exist concurrently. The sections where high-light overexposure and low-light underexposure occur are both marked as target sections, and the average brightness corresponding to these target sections is referred to as the target section brightness average.

More specifically, for all sections divided within the target image, there exist sections where the brightness mean exceeds the highest brightness threshold, and sections where the brightness mean is below the lowest brightness threshold. In the face of a situation where both high-light overexposure and low-light underexposure areas exist within the same image, merely increasing or decreasing exposure for target detection is not sufficient.

In detail, a preset number of high-low exposure images can be collected alternately based on the target section brightness average, thus obtaining images with more detailed brightness information based on these high-low exposure images to perform accurate target detection and achieve the target detection result.

The highest and lowest brightness thresholds have been described previously and will not be elaborated further here.

High-low exposure images include both high exposure and low exposure images. The preset number can be set according to actual conditions; for example, in one embodiment, six frames of high-low exposure images are alternately collected.

The preset number of frames of high-low exposure images are collected using an alternating exposure method, which specifically can be as follows:

1. One-step high exposure + one-step low exposure + average exposure;
2. Two-step high exposure + one-step low exposure + average exposure;
3. Two-step low exposure + one-step high exposure + average exposure;
4. Two-step high exposure + two-step low exposure + average exposure;
5. One-step high exposure + three-step low exposure + average exposure;
6. Two-step high exposure + three-step low exposure + average exposure;
7. Three-step high exposure + one-step low exposure + average exposure;
8. Three-step high exposure + two-step low exposure + average exposure;
9. Three-step high exposure + three-step low exposure + average exposure.

Typically, three-step cases are less common.

Based on the above, any combination of low exposure steps with high exposure steps can be used to collect the preset number of frames of high-low exposure images. Wherein, any low exposure steps include one-step low exposure, two-step low exposure, and three-step low exposure, and any high exposure steps include one-step high exposure, two-step high exposure, and three-step high exposure. In this embodiment, by obtaining the difference in brightness between the overall image brightness average and the brightness average of each section, and ensuring that the overall image brightness average is within a preset brightness range, and that there exist brightness differences exceeding the set brightness threshold, the second target detection mode is employed for target detection to obtain the target detection result.

2.2.3   Image Acquisition and Analysis Examples

Figure 2.4 shows a schematic diagram of the effects of a method for implementing wide dynamic range with multiple exposures in target detection.

As illustrated in Fig. 2.4, taking a desk as an example, the image "A" in Fig. 2.4 represents an image of the desk obtained under normal lighting conditions with a full-image balanced exposure, where it has been determined that the overall image brightness average falls within a preset brightness range. Image "C" represents an image of the same scene with local high-light exposure (that is, the section brightness

A: Normal lighting environment—balanced exposure of the entire image

B: Partial highlight environment - balanced exposure of the entire image

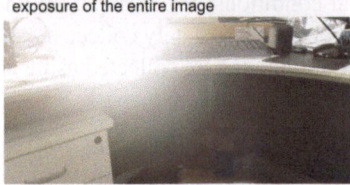

C: Partial highlight environment

D: Partial highlight environment—second-order low exposure (1)

E: Partial highlight environment - second-order low exposure (2)

Fig. 2.4 Images of the effects of a wide dynamic range implementation method for target detection

average is higher than the highest brightness threshold). In order to obtain the situation of target detection under the high-light environment in image "C" a two-step low exposure (second target detection mode) was performed on it. Two-step low exposure (1) can be seen in image "D" and two-step low exposure (2) can be seen in image "E" Finally, based on the image of two-step low exposure (2), a full-image balanced exposure is carried out to obtain image "B". It can be observed that performing a two-step low exposure on an image with local high-light exposure, both two-step low exposure (1) and two-step low exposure (2) can reveal image information under different brightness levels, such as the full view of the keyboard on the desk, the cup, papers, and the edge of the desk.

Compared to two-step low exposure (1), two-step low exposure (2) can also reveal image information under different brightness levels, for example, the words or markings on the paper in image "E" which are not detectable in the image of two-step low exposure (1).

2.2.4 Summary

Lastly, it is important to emphasize that the wide dynamic range implementation method for target detection provided by this design method, through the use of a common image sensor and without compromising frame rate, enables the capture of more brightness information in challenging lighting conditions such as under strong light sources, high-light and shadow areas, backlight, etc. This approach enhances target recognition in both high-light and excessively dark areas. The images do not require fusion or buffering, thereby reducing computation time and avoiding brightness information distortion that can arise from fusion.

It should also be noted that in this design method, software dynamically adjusts exposure parameters to produce multiple frames of differently exposed images, ultimately obtaining a single target detection result (image, text, and/or voice). This method can be applied to any camera and can be optimized according to the application scenario, with a focus on enhancing exposure control in target regions.

2.3 ISP Adapted for RGB-IR

Traditional CMOS image sensors use the Bayer pattern as the color filter array (CFA), where pixel color information is determined based on light of specific wavelengths. RGB-IR sensors represent a new type of sensor that adds an IR filter, capable of detecting infrared light wavelengths, into the RGB color model. With the addition of an IR camera, RGB-IR sensors can improve image quality in low-light environments. An RGB-IR image sensor is a new type of filter (Color Filter Array, or CFA) arrangement. Compared to the traditional Bayer pattern filter arrangement, it substitutes some of the color channel filters with infrared filters (IR), enabling simultaneous detection of both visible and non-visible light. Pixels in an RGB-IR image sensor have four types of filters: red (R), green (G), blue (B), and IR. Standard color image sensors have a color filter array (CFA) arranged according to the Bayer pattern. Currently, the CFA pattern for RGB-IR sensors has more variability, with 2×2 and 4×4 pixel arrays being quite common. A simple solution to add the IR channel to an RGB sensor is to reduce the number of red and blue in the Bayer pattern by half. RGB filters are selectively transmissive in the visible spectrum but are typically highly transmissive in the IR domain, thus RGB filters receive significant crosstalk from the IR channel.

Exposure of RGB pixels to IR radiation can reduce the image quality attributes of color images. As the IR portion increases, color accuracy, spatial resolution, signal-to-noise ratio, and dynamic range can all degrade. While acceptable for some computer vision applications, this can lead to serious local artifacts in smart facial recognition camera applications, affecting face recognition effectiveness and must be addressed. For color restoration of RGB-IR image sensors, the mainstream existing technical solution is to subtract the IR components that crosstalk into the RGB channels, and then proceed with traditional image processing algorithms. However, since R, G, and B pixels are actually superpositions of the IR channel and R, G, B channels and RGB filters have a high transmissivity of IR, saturation and noise are more likely to occur. When saturation occurs, directly subtracting the IR component can result in "faded" RGB colors, affecting image recognition.

Deriving high-quality RGB and IR images from an RGB-IR image can also be regarded as a demosaicing problem for non-traditional color filter arrays. Traditional demosaicing algorithms applied to RGB-IR sensors are impractical, first because each band has a different refractive index, leading to different focus between RGB and IR images; secondly, there is crosstalk introduced between RGB and IR channels; and furthermore, due to the presence of infrared light, the assumption of a linear relationship between the original image sensor RGB signal and the RGB image perceived by humans is no longer valid, thus traditional color correction algorithms are not supported. Therefore, both the demosaicing algorithm and the color correction algorithm for the RGB-IR image sensor need to be optimized.

In summary, RGB-IR image sensors face the following challenges when applied to smart facial recognition cameras:

1. Channel separation, decomposing the sensor's pixel values into pure R, G, B, and IR channels;
2. Eliminating crosstalk between channels;
3. Differential treatment of saturated and unsaturated pixels;
4. Optimization of the color correction algorithm.

2.3.1 Technical Solutions

This chapter proposes an advanced RGB-IR image color restoration and correction method in smart face camera applications to solve the problems of oversaturation and complex IR component removal operations and poor results in existing RGB-IR image color restoration, realizes the functions of dynamic adjustment of the weight of critical areas of RGB-IR images, adaptive anti-saturation processing in color restoration, and generation of a joint matrix for color restoration correction.

This chapter will provide a detailed explanation of the technical solution proposed in this chapter, including the advantages of this technical solution compared to traditional technical solutions and a detailed description of each link of this technical solution.

2.3.1.1 Comparison Between This Technical Solution and Traditional Technical Solutions

The general flow of image processing is shown in Fig. 2.5.

As shown in Fig. 2.5, the approach presented in this paper differs from traditional methods in the stages of raw image processing, color restoration, and color correction. Each stage will be specifically introduced based on these differences.

Raw Image Stage

In traditional approaches, video frames are processed uniformly without distinguishing between regions. We know that during the capture by a facial recognition camera, multiple images may be taken with suboptimal angles or brightness. These images, when processed together with other images, can negatively impact the final results. In our approach, video frames are pre-divided into multiple image blocks. These blocks are then evaluated to determine if they are critical regions, dividing

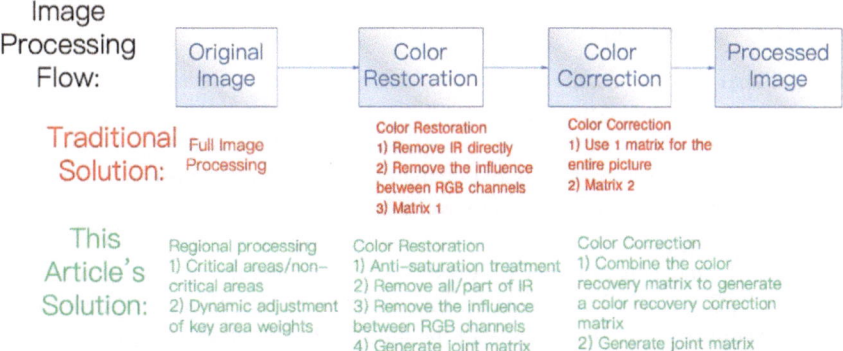

Fig. 2.5 General flow of image processing

them into critical and non-critical regions. Within critical regions, pixels are divided and layered for pixel traversal and histogram distribution statistics. Based on the distribution of the statistical histogram, weight values are determined.

Color Restoration Stage

In actual image processing, since R, G, and B pixels are superpositions of the IR channel and the respective RGB channels, and the RGB filters have a high IR transmissivity, saturation and noise are more likely to occur. When saturation happens, directly subtracting the IR component can cause the RGB colors to appear "faded". For example, if the G channel signal G0 is at a saturation level, subtracting the IR component at this time would result in a G1 signal below its correct value. After white balance processing, R1 and B1 would exceed G1, resulting in an excess of red and blue, manifesting as purple artifacts that affect image recognition. Thus, the traditional approach of directly removing IR is not feasible and anti-saturation processing is needed.

In our approach, adaptive anti-saturation is implemented. The proportion of the IR component and the saturation determination results dynamically implement differential processing for saturated and unsaturated pixels, as well as the calculation of the IR component proportion in unsaturated images. During saturation pixel determination, different thresholds are set based on the application scenarios of the smart facial recognition camera. The RGB of each pixel is compared with the set threshold; pixels exceeding the threshold are deemed saturated, while those equal to or less than the threshold are deemed unsaturated. Saturated and unsaturated pixels with different brightness levels are sent to different image processing paths.

In traditional approaches, there is only one matrix for color restoration, which is insufficient for RGB-IR smart facial recognition cameras. Combined with adaptive anti-saturation processing, for color restoration matrices that don't require anti-saturation, we use a 3×4 color restoration matrix to remove part or all of the IR component from the RGB channels. For color restoration matrices that do require anti-saturation, we use a 3×3 matrix, retaining the IR component in the RGB channels.

Color Correction Stage

In traditional approaches, the color restoration matrix and correction matrix are two separate matrices, so two matrix operations are required during the image processing. In our approach, the color restoration matrix and correction matrix are combined to create a joint matrix, requiring only one matrix operation during image processing.

In summary, the problems faced by RGB-IR image sensors in smart facial recognition camera applications, as mentioned earlier, are effectively resolved. In our approach, channel separation is achieved by decomposing the sensor's pixel values into pure R, G, B, and IR channels; the image is divided into critical and non-critical areas; by comparing RGB values with the threshold, critical area weight values are dynamically adjusted to generate comparison results; based on these results, it is determined whether anti-saturation processing is needed. Different color restoration correction joint matrices are used, which not only achieve adaptive anti-saturation processing but also precisely, efficiently, and conveniently eliminate the impact of the IR component on the image, optimizing the signal-to-noise ratio, enhancing the dynamic range, and based on critical area weights, generating the color restoration correction joint matrix to achieve image color restoration and correction.

2.3.1.2 Detailed Description of Each Part of This Technical Solution

Each part in the image processing process in this technical solution is shown in Fig. 2.6.

Based on Fig. 2.6, the image processing procedure of this technical solution is as follows:

1. Obtain an RGB-IR image and extract data for the R, G, B, and IR channels of each pixel; calculate the channel component values for the R, G, and B channels of each pixel, as well as the IR component value and proportion within each of the RGB channels.
2. Determine critical and non-critical areas.
3. Compare the RGB values of each pixel in the critical and non-critical areas with a set threshold; if the comparison result is greater than the set threshold, the pixel is considered saturated. if the comparison result is less than or equal to the set threshold, the pixel is considered unsaturated.

   Dynamically adjust the weight values of critical areas based on the number of saturated and unsaturated pixels in the critical and non-critical areas and the brightness distribution of the unsaturated pixels.
4. Based on the comparison results of the RGB values against the set threshold:

   In the case of unsaturated pixels, when the brightness is moderate, remove the entire IR component during color restoration, without anti-saturation processing.

   In the case of unsaturated pixels, when the brightness is low, partially remove the IR component during color restoration, with partial anti-saturation processing.

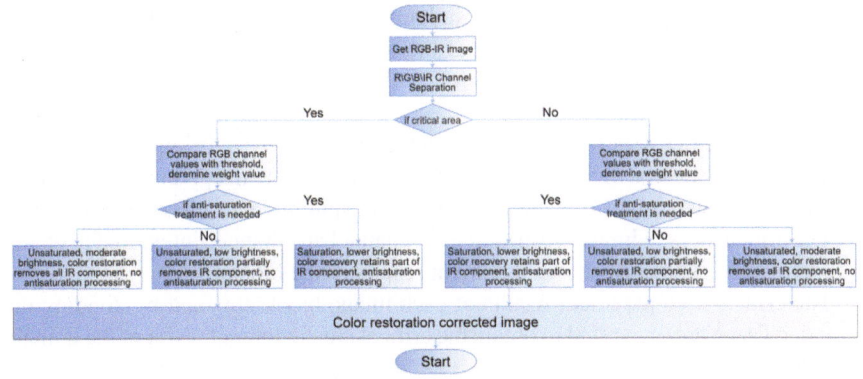

Fig. 2.6 Various parts in the image processing process

In the case of saturated pixels, retain the IR component during color restoration, with anti-saturation processing.

5. Depending on the critical area/non-critical area and whether anti-saturation processing is required, different color restoration correction matrix coefficients are generated to complete color restoration correction.

In the whole process, extracting the data of R channel, G channel, B channel and IR channel of each pixel, which is R\G\B\IR channel separation, is a relatively mature technology. The following focuses on the determination of critical areas and non-critical areas, adaptive anti-saturation, color restoration correction matrix and typical applications.

2.3.2   Determination of Critical Areas and Non-critical Areas

The judgment of critical areas and non-critical areas includes:

- Obtain the video frame to be determined and 4 adjacent video frames; the video frame is pre-divided into multiple image blocks; determine whether the image block in the video frame is a critical area; if the image block in the video frame is a critical area, then the image block Perform VST transformation based on known critical area parameters and then perform channel splitting and packaging based on Baver format; if the image blocks in the video frame are non-critical areas, channel splitting and packaging are directly performed on the image blocks based on Bayer format;
- After channel splitting and packaging of image blocks based on Bayer format, the R channel image block sequence, Gr channel image block sequence, B channel image block sequence and Gb channel image block sequence are obtained;
- According to whether the image block before channel splitting is a critical area, the image block sequence of each channel obtained by splitting and packaging the critical area image blocks is sent to the VST domain critical area judgment network to obtain the critical area judgment results of each channel, and each The critical area judgment results of the channel are subjected to VST inverse transformation and channel integration to obtain the critical area judgment results of each image block;
- The image block sequences of each channel obtained by splitting and packaging the non-critical area image blocks are respectively sent to the pixel domain critical area judgment network to obtain the critical area judgment results of each channel. The critical area judgment results of each channel are channel-integrated to obtain each image. The critical area judgment results of the block;
- The critical area judgment results of each image block obtained by dividing the video frame are integrated to obtain the critical area judgment result of the domain video frame.

2.3.3 Adaptive Anti-saturation

For unsaturated pixels with moderate brightness, the RGB channel can remove all IR components. When the RGB brightness is sufficient, the smart face camera can use high-quality RGB images to obtain clear and correct face images. Based on this purpose, the obtained color restoration correction matrix can eliminate the infrared light information in the R, G, and B component images and complete color restoration.

For unsaturated pixels with low brightness, it is only necessary to remove part of the IR component of the RGB channel and retain part of the IR component in the RGB channel. This is because when the brightness of RGB is insufficient, the smart face camera only uses RGB images and cannot obtain subsequent results. For face images processed by image processing, the presence of IR components can improve the brightness of the image to a certain extent. Based on this purpose, the obtained color restoration correction matrix can eliminate part of the infrared light information in the R, G, and B component images and complete color restoration.

For saturated pixels, retain part of the IR component of the RGB channel. The reason why the IR component is retained in the RGB channel is because when the RGB is saturated, removing the IR component at this time will cause the value of some RGB channels to be lower than the correct value. In the subsequent white balance When processed by the algorithm, artifacts will result. Artifacts are particularly influential in smart face cameras. Based on this purpose, the obtained color restoration correction matrix can retain the infrared light information in the R, G, and B component images to achieve the anti-saturation effect.

Regardless of whether there are saturated pixels in the input image, it can be processed correctly and a high-quality RGB image can be obtained. Dynamic processing of saturated pixels and unsaturated pixels of different brightness allows the smart face camera to finally obtain the ideal image.

2.3.4 Color Restoration Correction Matrix

2.3.4.1 Color Restoration Correction Matrix for Unsaturated Pixels

For unsaturated pixels, the decision to remove either the entire IR component or a part of it is based on the brightness level and the proportion of the IR component. Drawing on the spectral response curve of the CMOS image sensor, settings for the IR component removal ratio, and critical area weights, two color restoration correction matrices (3×4 matrices) can be obtained. These two matrices can respectively achieve the removal of all IR components and part of the IR components from the RGB channels, eliminating the crosstalk between the RGB channels while simultaneously accomplishing color correction. Both matrices are 3×4 matrices, with the only difference being the coefficients. They are detailed as follows:

Critical Areas

Without the need for anti-saturation processing, the decision is made to remove either the entire IR component or a part of it from the RGB channels based on the brightness level and the proportion of the IR component; To determine the first color restoration correction matrix for critical areas, based on critical area weights and the objective of removing all IR components from the RGB channels, the matrix is obtained as follows:

$$[R_2\, G_2\, B_2] = \left[R_r^{3\times4} R_g^{3\times4} R_b^{3\times4} G_r^{3\times4} G_g^{3\times4} G_b^{3\times4} B_r^{3\times4} B_g^{3\times4} B_b^{3\times4} R_{ir}^{3\times4} G_{ir}^{3\times4} B_{ir}^{3\times4} \right]$$
$$\times [R\,G\,B\ IR]$$

According to the weight of the critical area and the purpose of removing part of the IR components in the RGB channel, a second color restoration correction matrix of the critical area is obtained. The second color restoration correction matrix of the critical area is different from the first color restoration correction matrix only in coefficients.

Non-critical Areas

When anti-saturation processing is not required, determine whether to remove all IR components or part of the IR components in the RGB channel according to the brightness value and the proportion of the IR components.

According to the weight of the non-critical area and the purpose of removing all IR components in the RGB channel, the first color recovery correction matrix of the non-critical area is obtained:

$$[R_2\, G_2\, B_2] = \left[R_r^{3\times4} R_g^{3\times4} R_b^{3\times4} G_r^{3\times4} G_g^{3\times4} G_b^{3\times4} B_r^{3\times4} B_g^{3\times4} B_b^{3\times4} R_{ir}^{3\times4} G_{ir}^{3\times4} B_{ir}^{3\times4} \right]$$
$$\times [R\,G\,B\ IR]$$

According to the weight of the non-critical area and the purpose of removing part of the IR components in the RGB channel, a second color restoration correction matrix for the non-critical area is obtained. The second color restoration correction matrix for the non-critical area is different from the first color restoration correction matrix only in coefficients.

2.3.4.2   Color Restoration Correction Matrix for Saturated Pixels

For saturated pixels, it is necessary to retain the IR component of the RGB channel. According to the spectral response curve of the CMOS image sensor and the critical area weight, a color recovery correction matrix (3×3 matrix) can be obtained. This color recovery correction matrix can retain the IR component of the RGB channel. Crosstalk between RGB channels is removed while color correction is completed. The matrix is a 3×3 matrix, as follows:

Critical Areas

The color restoration and correction of saturated pixels in the critical area is performed through the third color restoration correction matrix in the critical area. The third color restoration correction matrix in the critical area is:

$$[R_1 \, G_1 \, B_1] = \left[R_r^{3\times3} R_g^{3\times3} R_b^{3\times3} G_r^{3\times3} G_g^{3\times3} G_b^{3\times3} B_r^{3\times3} B_g^{3\times3} B_b^{3\times3} \right] \times [R \, G \, B]$$

Non-critical Areas

The color restoration and correction of saturated pixels in non-critical areas is performed through the third color recovery correction matrix in non-critical areas. The third color recovery correction matrix in non-critical areas and the third color recovery correction matrix in critical areas are both 3 × 3 matrices, but the coefficients are different.

2.3.5 Typical Applications

Smart Facial Recognition Camera Capturing Faces in Strong Light, Resulting in Overexposed Faces

When faces are under strong light, images captured by the smart facial recognition camera show overexposure in the facial area, which is the critical region, while the background area, the non-critical region, appears relatively darker compared to the face. Assume that the G channel of the critical region is overexposed, while the R and B channels are not. If we follow the general practice of RGB-IR image processing, which is to first remove the IR component from the RGB channels and then perform white balance, the resulting pixels will have R and B channels reaching the specified pixel values, but the G channel will be below the specified value. In this case, what should be a white image will appear with purple artifacts.

After analyzing the image captured by the smart facial recognition camera, it is found that the saturated pixels are concentrated in the critical region, and the overall brightness of the critical region is greater than that of the non-critical region. At this time, the weight proportion of the critical region is about 80%. Color restoration correction matrices are generated based on this weight for both saturated and unsaturated pixels, completing the color restoration and correction. The processed facial images retain the IR component in saturated pixels, without the appearance of artifacts. With a higher weight in the critical region, the color correction effect is more conducive to subsequent facial image processing algorithms.

Smart Facial Recognition Camera Capturing Faces in Indoor Low Light, with Low Brightness in Both Face and Background Areas

After analyzing the image captured by the smart facial recognition camera, there are no saturated pixels, and the overall brightness of the unsaturated image is low, with no significant contrast between the critical and non-critical areas. In this case, the weight proportion of the critical region is about 50%. Color restoration correction matrices are generated based on this weight for unsaturated pixels of different brightness levels, completing the color restoration and correction. In the processed facial images, only part of the IR component is removed for unsaturated pixels with low brightness, and the retained IR component is beneficial for subsequent image processing algorithms. The weight of the critical and non-critical areas is similar, leading to better uniformity in the corrected image.

Smart Facial Recognition Camera Capturing Faces Indoors Against Strong Backlight, with Dark Faces and Overexposed Background

After analyzing the image captured by the smart facial recognition camera, it is found that the saturated pixels are concentrated in the non-critical area, and the overall brightness of the critical region is less than that of the non-critical region. At this time, the weight proportion of the critical region is about 20%. Color restoration correction matrices are generated based on this weight for both saturated and unsaturated pixels, completing the color restoration and correction. In the processed facial images, the saturated pixels in the background retain the IR component, without the appearance of artifacts, not affecting subsequent image processing. With a lower weight in the critical region, the color correction effect appears more natural, and compared to the background, it does not stand out abruptly.

2.4   Task Scheduling Mechanism of Active Camera

Embedded real-time task scheduling systems play a crucial role in embedded systems and real-time applications, particularly in scenarios where real-time requirements must be met. Here are some key functions of real-time task scheduling systems:

1. Ensuring Timely Task Completion: The main goal of real-time task scheduling systems is to ensure that real-time tasks within the system are completed within their specified time frames. For certain applications, such as real-time control systems and embedded real-time image processing, timely task completion is critical; otherwise, it could lead to decreased system performance or serious safety issues.
2. Improving System Response Speed: Real-time task scheduling systems allocate and manage system resources efficiently, allowing higher-priority tasks to be executed more quickly. This is vital for systems that need to respond rapidly to external events, such as autonomous driving vehicles and robotic controls.

3. Reducing System Latency: In some real-time applications, such as audio–video processing and communication systems, system latency is an important performance metric. Real-time task scheduling systems effectively schedule tasks to reduce waiting times between them, thereby decreasing the overall system latency.
4. Enhancing Resource Utilization Efficiency: Real-time task scheduling systems dynamically allocate system resources to ensure that high-priority tasks are executed preferentially, improving the effective utilization of system resources. This is especially important in resource-constrained environments like embedded systems.
5. Dealing with Uncertainty: Real-time systems often face challenges due to changes in the external environment and uncertainties in task execution times. Real-time task scheduling systems need to be robust, capable of adapting to fluctuations in task execution within unpredictable environments, to ensure the system still meets real-time requirements.

Overall, real-time task scheduling systems provide reliable foundational support for real-time applications by ensuring real-time performance, effective resource management, and task scheduling, making the system more stable, responsive, and efficient in resource use.

Especially in the field of embedded camera systems, task scheduling and resource management are among the key technical challenges. Embedded camera systems integrate face image capture and face recognition algorithms, embedding these functions into a compact embedded platform aimed at efficiently coordinating image capture, image processing, and part of the face algorithms. However, under the same hardware conditions, the method of task scheduling in the camera is crucial for improving camera response times and resource utilization efficiency.

Embedded camera platforms often face challenges due to limited hardware resources, especially when considering factors such as energy consumption, cost, and size. These platforms are typically equipped with only a single-core dedicated computing acceleration hardware, such as a single-core GPU or DSP.

To address this issue, modern embedded camera systems need to adopt smarter, more adaptive task scheduling methods. This involves designing flexible scheduling logic that can dynamically calculate the priorities of various tasks and dynamically adjust priorities based on real-time requirements and computation time differences. This new scheduling approach can better adapt to the unique hardware architecture of embedded camera platforms, improving overall system performance, ensuring better camera response times, and resource utilization efficiency. This remains one of the ongoing research and development directions in the field of embedded camera systems.

Traditional Embedded System Scheduling Methods

Traditional embedded software task scheduling often employs a first-come-first-served preemptive method, where the task that first secures computing resources will monopolize these resources until its completion. The advantage of this method is its simplicity, and when computing resources are abundant, it can ensure a certain level of computational efficiency and resource utilization. However, when hardware resources are limited, particularly on many low-cost embedded camera platforms, due to considerations such as energy consumption, cost, and size, these platforms are typically equipped with only a single-core dedicated computing acceleration hardware, such as a single-core GPU or DSP.

In such cases, simply adopting a first-come-first-served preemptive scheduling approach may cause the system to wait for tasks with low real-time requirements, long computation times, and low priority to finish. For example, some image processing tasks may require longer computation times while having relatively low real-time requirements. Such tasks could occupy computing resources for an extended period, preventing the system from promptly executing high-priority tasks with high real-time requirements and short computation times, like facial recognition or real-time image tracking. This could lead to increased system response times, reduced resource utilization, and thus affect the user experience, causing issues such as frame drops and delays.

To address this issue, more intelligent and adaptive task scheduling methods are needed to better accommodate the unique hardware architecture of embedded camera platforms. New scheduling methods must take into account factors such as real-time requirements, computation times, and task priorities to optimize the scheduling of computational tasks and improve overall system performance. This involves designing flexible scheduling logic capable of dynamically calculating each task's priority and adjusting priorities dynamically based on real-time requirements and computation time differences.

Real-Time Task Dynamic Scheduling Method

To address the range of issues presented in current methods, this paper proposes an innovative task scheduling approach. The goal is to optimize the scheduling of computing tasks by taking into account a combination of factors: real-time requirements, computation times, and priorities, thereby enhancing the overall performance of the entire system. This proposed scheduling method is targeted and can effectively resolve the various shortcomings of traditional preemptive scheduling methods under conditions of resource scarcity.

2.4.1 Overall Framework

Firstly, by intelligently sensing the surrounding environment and dynamically analyzing real-time requirements, the system can more intelligently respond to the urgency of various tasks. The trigger frequency and real-time requirements of different tasks vary, and this scheduling method can flexibly adjust task priorities based on the number of requests per second and computational time requirements of each task, ensuring timely responses to tasks with high real-time needs, thus avoiding issues like image frame dropping and jitter.

Secondly, considering that each computational task requires different amounts of time, the system dynamically allocates priorities to more effectively handle tasks with shorter computation times, aiming to release computing resources sooner and improve the overall system response speed. This differentiated task scheduling strategy allows the system to operate more efficiently and maximize the potential of hardware resources.

The dynamic adjustment of priority queues and the flexible switching of interrupting tasks enable the system to better adapt to different real-world application scenarios and workloads. This flexibility provides the system with stronger adaptability, allowing it to maintain efficient and stable operation in various environments.

Additionally, the introduction of a Real-Time Operating System (RTOS) is another means of enhancing task scheduling efficiency. RTOS is specifically designed for real-time applications and offers more refined task scheduling and management mechanisms. Using RTOS can better meet the needs of scenarios with stringent real-time requirements, ensuring that the system completes tasks within the specified time. When choosing RTOS, its compatibility with the hardware platform and its impact on system performance should be considered.

On single-core embedded camera platforms, task segmentation and parallel processing are also important strategies. By dividing tasks into smaller subtasks and executing them in parallel, hardware resources can be more fully utilized. This has clear advantages for handling complex tasks and improving system throughput. However, task segmentation requires a well-designed dependency relationship between tasks to ensure the effectiveness of parallel execution.

Adopting a dynamic scheduling strategy, which continuously monitors system load and real-time requirements of tasks, allows for dynamic adjustment of task scheduling to adapt to different workloads. This method can more flexibly respond to changes during system operation and maximize the use of limited hardware resources.

In practical applications, task scheduling performance is not only related to the algorithm itself but also closely tied to the characteristics of the hardware platform. Therefore, when designing task scheduling strategies, the features of the hardware must be fully considered, including the quantity, speed, and sharing methods of computational resources. Additionally, dependencies between tasks need to be considered to avoid issues such as data races and deadlocks.

The scheduling method consists of scheduling logic, priority queues, and underlying computational resources. It is designed to overcome the shortcomings of traditional preemptive scheduling methods and achieve smarter and more flexible task management.

2.4.1.1 Scheduling Logic

In embedded camera systems, the scheduling logic, as the core component of task scheduling, plays a pivotal role. Its responsibilities include receiving a variety of computational task requests from the facial recognition camera's upper-layer applications, which encompass aspects such as ISP image processing, full-image face detection, face tracking, and face recognition. Each task has its unique computational requirements and real-time demands, so the scheduling logic must work closely with the upper-layer applications to deeply understand the specifics of the tasks, providing key information for subsequent priority calculation and task queue management.

The scheduling logic needs to accurately acquire task requests from upper-layer applications. This involves a detailed analysis and interpretation of different computational tasks' characteristics. For example, an ISP image processing task may require completion at a rate of 30 frames per second, while the full-image face detection task may be requested 2–3 times per second, and the face tracking and recognition tasks each have their own trigger frequency. Understanding the frequency of task triggers helps to assess their real-time requirements, thus providing a basis for subsequent priority calculations. This step requires the scheduling logic to be sensitive and understanding of various task types to ensure accurate acquisition of key parameters of task requests.

A close communication mechanism must be established between the scheduling logic and upper-layer applications to gather more detailed information about the tasks. This includes the specific computational requirements of the task, the occupation of hardware resources, and the specific real-time demands. Through close collaboration with the upper-layer applications, the scheduling logic can gain a deeper understanding of each task's unique needs, providing more comprehensive information for subsequent priority calculation and resource allocation. In this process, the scheduling logic may need to communicate with the facial recognition camera application developers to obtain deeper task characteristics, thereby better serving the optimization of overall system performance.

After receiving the task requests and understanding their characteristics, the scheduling logic needs to perform a reasonable priority calculation and task queue management for these tasks. Different types of tasks should be assigned different priorities based on factors such as their real-time requirements, trigger frequency, and computation time. The scheduling logic must possess intelligent scheduling capabilities to ensure that high-priority tasks receive timely responses, thereby improving the overall performance of the system. At the same time, the management of the task queue is crucial—ensuring that tasks enter the queue in an orderly manner according to priority levels, so as to maximize the execution of high-priority tasks when computing resources are limited. These series of operations require the scheduling logic to have good algorithm design and implementation to effectively enhance the overall performance of the embedded camera system.

2.4.1.2   Priority Queue

The design of the priority queue plays a key role within the entire scheduling method, with its flexibility and intelligence providing significant support for performance enhancement in embedded camera systems. The order of tasks in the queue is not static but dynamically adjusted according to task priorities, providing the system with high adaptability and flexibility. As one of the core components of the scheduling method, the design of the priority queue enables the system to maximize overall performance by flexibly adjusting the execution order based on key factors such as the real-time requirements and computation times of the current tasks.

A dynamically adjusted priority queue allows the system to intelligently address the needs of different tasks. In this process, the scheduling logic acts as the decision-maker, making real-time adjustments to the priority queue based on the dynamically calculated priorities of each task. This intelligent decision-making mechanism enables the system to flexibly switch between different tasks, prioritizing those with higher real-time requirements, and thus increasing the overall system's response speed. With real-time dynamic adjustments, the priority queue is no longer a static execution order but changes based on the urgency and complexity of tasks, providing better adaptability for the system.

The dynamically adjusted priority queue is sensitive to computation times, aiding in the rational allocation of computing resources. Since different tasks require varying computation times, the dynamically adjusted priority queue takes this factor into account. The system can better utilize computing resources by allocating more resources to tasks that can be completed more quickly, thereby enhancing the overall efficiency of task execution. This sensitivity to computation times allows the system to allocate resources more intelligently under limited resources, optimizing overall performance.

Finally, the real-time adjustment mechanism of the priority queue offers the system more agile responsiveness. The scheduling logic can quickly assess the urgency of current tasks and adjust the queue based on real-time calculations, allowing the system to swiftly adapt to different workloads and task demands. This agile responsiveness helps maintain the efficiency and stability of the system across different scenarios, ensuring that tasks are successfully completed even under high real-time requirements.

In summary, the dynamically adjusted priority queue, as a highlight of the scheduling method, provides a more efficient and flexible way of executing tasks for embedded camera systems through intelligent priority management, sensitivity to computation times, and agile responsiveness. This drives continuous improvements in the overall performance of the system.

2.4.1.3 Underlying Computing Resource Management

The underlying computational resources, as the executors of task scheduling, directly influence the completion speed of tasks and the overall responsiveness of the embedded camera system. This layer of hardware resources, typically including single-core dedicated computational acceleration hardware such as GPUs or DSPs, plays a crucial role in the execution of tasks in the camera system. Therefore, the scheduling method must consider the characteristics of these underlying computational resources in its design to maximize task execution efficiency.

The full utilization of underlying computational resources is vital to the completion speed of each task. As these hardware resources take on the leading role in computation within the camera system, the scheduling logic needs to allocate tasks cleverly to ensure that each task can fully tap into the potential of the underlying computational resources. By deeply understanding the computational capabilities and characteristics of the underlying resources, the scheduling method can make more intelligent decisions regarding task allocation, allowing each task to be completed in the shortest possible time, thus enhancing the overall task execution efficiency of the system.

The status of the underlying computational resources must be fully considered in the decision-making process of task scheduling. Since the underlying computational resources are often limited, there may be multiple tasks competing for the same resource. The scheduling logic should have the ability to monitor and assess the status of the underlying resources in real-time to better inform the scheduling decisions. With a comprehensive understanding of the computational resource status, the scheduling method can avoid overuse and wastage of resources, thereby allocating computational resources more effectively and ensuring the stability and efficiency of the overall system.

The scheduling method must flexibly respond to dynamic changes in the underlying computational resources during task execution. As the status of hardware resources can change over time, the scheduling logic should be capable of dynamically adjusting the order of task execution. Such a mechanism allows the system to better adapt to the resource conditions at different points in time, thereby enhancing the overall performance of the system. Through real-time dynamic adjustments to the order of task execution, the scheduling method can more flexibly respond to changes in hardware resources, ensuring that the system maintains efficient operation under various workloads.

The full utilization of underlying computational resources, thorough consideration of resource status, and flexible response to dynamic changes present key challenges for the scheduling method in embedded camera systems. By deeply understanding and comprehensively considering these factors, the scheduling method can manage the underlying computational resources more intelligently and efficiently, thereby driving improvements in the overall system performance.

2.4.1.4 Smart Camera Task Priority Algorithm

When establishing the priority of computing tasks for an intelligent camera system, it is necessary to consider multiple factors to ensure the system operates efficiently with limited resources.

Trigger frequency is a key factor to consider. Different tasks have varying request frequencies for the system. For instance, at the speed of 30 frames per second for image processing, the ISP image processing task requests 30 times per second, full-image face detection tasks request 2–3 times per second, face tracking tasks request 30 times per second, while face recognition tasks request once per second. This not only reflects the real-time requirements of tasks but also provides a reference for the allocation of priorities. Tasks with higher real-time requirements have more urgent demands for system resources and, thus, might be assigned higher priorities in the priority queue.

The computation time required for different tasks must be taken into account. Taking ISP image processing, full-image face detection, face tracking, and face recognition tasks as examples, they each occupy different computational resources and have various computation times. The ISP image processing task needs to complete computations within every 30 ms to ensure processing is done before the next frame arrives, avoiding issues such as frame dropping and jitter. Consequently, considering the length of computation time, task priority might also be adjusted accordingly. To maximize overall system performance, we tend to assign higher priority to tasks that are relatively short in computation time and can be completed quickly, ensuring they are executed fully within a limited time frame.

Beyond trigger frequency and computation time, the cost of interrupting the current executing task also needs to be considered. Interrupting a current task may involve saving the current state, loading new task code and data, which introduces additional overhead. In this regard, we need to appropriately adjust the weight of the switching cost to avoid frequent task switching that could degrade system performance. In some cases, even if a task has a high frequency of triggers and real-time requirements, if the cost of interrupting the task is high, more cautious decision-making might be required when switching tasks to maintain overall system stability.

The formula for calculating task priority is:

$$P = a \cdot R + b \cdot \left(\frac{Tw}{Tr}\right) + c \cdot S$$

Which P represents the task priority, R represents the real-time nature of the task, Tw represents the task waiting time, Tr represents the remaining time of the task, S represents the cost of switching tasks. In the formula, a, b, and c are weights, which can be adjusted according to system characteristics and needs.

This priority calculation formula takes into account multiple factors such as the real-time nature of the task, waiting time, remaining time, and switching cost. Specifically, the $a \cdot R$ part represents the real-time nature of the task, which is more important for tasks with high real-time requirements. By adjusting the value of a, the weight of real-time performance in the overall priority can be flexibly adjusted, so that real-time tasks can be executed with higher priority in systems with high real-time requirements.

On the other hand, $b \cdot \left(\frac{Tw}{Tr}\right)$ takes into account the ratio of the task's waiting time and remaining time. When a task has a longer waiting time or a shorter remaining time, its priority will be increased accordingly. This helps avoid low-priority tasks not being executed for a long time and improves the overall response speed of the system.

Finally, $c \cdot S$ considers the cost of switching tasks, including time overhead such as saving the scene and loading new task code and data. By adjusting the value of c, the impact of switching costs on the overall priority can be flexibly controlled to avoid system performance degradation caused by frequent task switching.

Such a task priority calculation formula provides the system with a flexible and comprehensive task scheduling strategy, enabling it to intelligently adjust the order of task execution based on specific circumstances, and thereby improving overall system performance and responsiveness.

In summary, establishing the priority of computing tasks requires a comprehensive consideration of factors such as trigger frequency, computation time, and switching costs. This multidimensional approach helps to achieve reasonable allocation of system resources, improving overall system performance and response speed. Through meticulous management of task priorities, embedded camera systems can better handle the challenges of various computing tasks, ensuring users have a smooth and efficient experience. In the future, as technology continues to innovate and hardware is upgraded, the optimization of task scheduling strategies will remain an important direction of research to meet the growing computational demands.

2.4.2   Application of Dynamic Scheduling Method in Face Camera

By coordinating the scheduling logic, priority queue, and underlying computational resources, intelligent scheduling of diversified computing tasks in facial recognition camera systems is achieved. By dynamically calculating the priority of tasks, their real-time requirements, and reasonably managing the underlying computational resources, the system can complete tasks more efficiently, enhancing the camera's response speed and overall performance.

2.4.2.1   Face Camera Task Scheduling Requirements

In the context of facial recognition camera systems, upper-layer applications issue a variety of computational task requests to the scheduling logic, providing a rich and diverse range of functional demands for the system's collaborative work. These tasks encompass multiple aspects such as ISP image processing, full-image face detection, face tracking, and face recognition, each with unique computational requirements and real-time demands, presenting the scheduling logic with complex and varied challenges in task scheduling and priority management [5, 6].

ISP image processing tasks hold an important position within the camera system. This task requires real-time processing of images to meet the requirements for high-quality imagery. Given the high real-time demands of ISP image processing tasks, the scheduling logic must consider their trigger frequency when calculating task priorities, which typically make 30 requests per second. This also implies that the task may enjoy a higher priority within the priority queue to ensure computations are completed within the short duration of each image frame, avoiding issues such as frame dropping and jitter.

Full-image face detection tasks are also a crucial component of facial recognition camera systems. This task has relatively high computational complexity and needs to be completed at a frequency of 2–3 times per second. Due to its lower trigger frequency, the scheduling logic might place it lower in the priority queue. However, considering its longer computation time, the scheduling logic also needs to collectively consider the priorities of different tasks in the decision-making for task switching to maximize the overall performance of the system.

Face tracking and face recognition tasks strike a balance between real-time demands and computation time. The former needs to be completed at a frequency of 30 times per second, while the latter operates at a frequency of once per second. When calculating task priorities, the scheduling logic must consider not only the real-time requirements but also the differences in computation time. In the priority queue, these two tasks may occupy a medium position, requiring the scheduling logic to intelligently and flexibly adjust according to the actual situation.

2.4.2.2 Main Tasks of Face Camera

The facial recognition camera, as an intelligent system integrated with multiple advanced technologies, requires effective scheduling of various tasks to ensure efficient system operation. The following are the main tasks that need to be scheduled in a facial recognition camera:

- Camera Control Algorithms: Camera control algorithms involve a wide range of technologies, including region-based exposure enhancement and multi-frame wide dynamic exposure. These algorithms need to dynamically adjust camera parameters in different shooting scenarios to ensure the quality and clarity of the images. Task scheduling is required to switch and optimize these camera control algorithms in real time based on different environmental conditions and application demands.
- Image Processing and Enhancement: Image processing and enhancement tasks include generic ISP (Image Signal Processing), RGBIR (Infrared–Visible Light) anti-saturation algorithms, among others. These algorithms are used to improve the quality of images, enhancing contrast and color saturation. Task scheduling is necessary to apply these algorithms appropriately to ensure that captured images have the best effect under various lighting conditions.
- Full-Image Face Detection Algorithms: Full-image face detection is one of the key tasks in facial recognition cameras. The algorithm needs to quickly and accurately detect the position of faces throughout the entire image. Task scheduling must optimize the execution order and frequency of the algorithms to ensure that full-image face detection can function effectively in various scenarios.

- Face Tracking Algorithms: Face tracking algorithms perform real-time tracking of faces based on previously detected face positions. Task scheduling should reasonably arrange the priority of face tracking to ensure stable tracking of moving faces, and dynamically adjust algorithm parameters when necessary.
- Facial Landmark Localization Algorithms: Facial landmark localization algorithms are used to identify key feature points on the face, such as eyes, nose, mouth, etc. Task scheduling needs to decide when to execute the landmark localization algorithms based on specific application scenarios to improve the accuracy of facial features.
- Face Capture Algorithms: Face capture algorithms are used to capture clear facial images under specific conditions. Task scheduling needs to ensure that this algorithm can respond promptly when face capture is required, guaranteeing the acquisition of high-quality facial images.
- Face Recognition Algorithms: Face recognition algorithms are used for comparing and identifying registered faces. Task scheduling must arrange the timing and frequency of face recognition to meet real-time requirements and dynamically adjust based on demands.
- Feature Comparison Algorithms: Feature comparison algorithms are used to compare the collected facial features with those in the database. Task scheduling needs to ensure that the algorithm can execute efficiently when feature comparison is required, achieving accurate face matching.

Task scheduling is crucial for the overall performance and user experience of the facial recognition camera system. A rational scheduling strategy can intelligently organize tasks based on real-time demands and system load, enhancing the system's efficiency and reliability.

2.4.2.3 Application Scenario Example: Face Detection

In facial recognition camera systems, face detection is a key algorithm. However, sometimes the system may encounter issues where a low-priority, slow full-image face detection algorithm results in a lower frequency of full-image face detection. The low-priority, slow algorithms may lead to slower detection speeds due to complex computations and large-scale image processing, thereby reducing the frequency of full-image face detection. In such cases, the system might miss some quickly appearing faces, affecting real-time performance. To compensate for this deficiency, we can adopt an intelligent and efficient strategy by introducing a high-priority, fast face tracking algorithm to improve the system's real-time performance and the frequency of full-image face detection [7].

The high-priority face tracking algorithm can perform rapid face detection based on the results of the most recent full-image detection. Thanks to its speed, it can process images in a short time, achieving real-time face detection effects. By introducing face tracking on top of the full-image detection algorithm, we can compensate for the shortcomings of a lower frequency of full-image detection, allowing the system to more flexibly handle face detection demands in various scenarios.

The working principle of the face tracking algorithm is to perform a rapid, small-scale detection of faces in the image based on the results of the most recent full-image detection. This way, even if the full-image detection algorithm cannot be executed frequently due to its low priority, the high-priority face tracking algorithm can complete the detection task quickly, improving the system's response to real-time requirements. This strategy is particularly suitable for scenarios that require quick capturing of changes, such as moving persons or rapidly changing environments.

By introducing a high-priority, fast face tracking algorithm into the system, we can effectively compensate for any real-time deficiencies that may be caused by a low-priority, slow full-image face detection algorithm. This intelligent scheduling strategy allows for flexible adjustment of the execution order of algorithms based on their priorities and computational demands, enhancing the overall system performance and the effectiveness of face detection.

2.4.2.4 Application Scenario Example: Exposure Control in Low Light Environment

The key role of camera control algorithms in scenarios of low-light adaptability cannot be overlooked [8]. In situations of darkness or insufficient lighting, to ensure that the facial recognition camera system operates normally and provides high-quality images, camera control algorithms become a critical technology. In such scenarios, we need to flexibly respond to changes in lighting conditions through an intelligent task scheduling mechanism.

We can adjust the real-time computational performance of the camera control algorithms. This aspect involves exposure enhancement and multi-frame wide dynamic range algorithms, which are essential for improving image quality under low-light conditions. By enhancing the real-time computational performance of these algorithms, the system can respond more swiftly to lighting changes, ensuring that the camera captures clear, well-lit facial images even in extreme conditions.

To ensure that these camera control algorithms are given higher priority, an intelligent task scheduling system is needed. In low-light environments, the system needs to invoke these algorithms more frequently to adapt to the challenges of insufficient lighting. By dynamically adjusting the priorities of tasks, the system can prioritize tasks related to camera control, thus enhancing the performance of the camera system in such scenarios.

Adjusting the priority of camera control algorithms not only helps improve image quality but also contributes to the stability of the entire facial recognition camera system. By strengthening the scheduling and resource allocation for camera control tasks in low-light environments, the system can adapt more reliably to complex lighting conditions, thereby providing superior face recognition results. This intelligent task scheduling mechanism is a key factor in ensuring outstanding performance in all types of environments.

2.4.2.5   Application Scenario Example: Non-sensory Face Recognition

Frictionless facial recognition demands higher speeds for identifying faces, necessitating the recognition of faces at a faster rate. In scenarios requiring high-speed facial recognition, adjusting task priority weights is a key strategy to enhance system performance. Specifically, by increasing the weight of parameter (b), which directly relates to the computational time and real-time requirements of tasks, such adjustments can significantly impact the system's response speed to tasks like face detection, facial landmark localization, capture, and facial recognition. This optimization strategy is expected to play a positive role in applications where high-speed facial recognition is demanded, such as access control systems and facial payment systems [9].

For embedded real-time task scheduling systems, we have delved into the pivotal role of task scheduling and its indispensable function in improving overall system performance. Through the collaborative efforts of scheduling logic, priority queues, and underlying computational resources, embedded systems can more intelligently cope with diversified computational tasks within the constraints of limited hardware. This ensures that high-priority real-time tasks are executed promptly, thus meeting stringent expectations for system response speed, task completion times, and other real-time requirements.

2.4.3   Summary

This chapter provides an in-depth analysis of the design principles of real-time task scheduling systems. Starting from factors such as task trigger frequency, computation time, priority, and resources, it proposes a method that dynamically adjusts task scheduling based on real-time requirements, computational time, and priority. This flexible and intelligent scheduling mechanism effectively enhances the system's adaptability to different tasks while maximizing resource utilization efficiency.

In specific application scenarios, we also explored the task scheduling strategies in facial detection settings, under varying lighting conditions, in scenarios with high demands for facial recognition speed, and other situations requiring real-time performance. By properly adjusting the weights of task priorities, the system can better adapt to a variety of application needs, ensuring that embedded cameras perform well in various complex environments.

In summary, as the core engine of embedded systems, real-time task scheduling systems not only improve the real-time performance of the system but also provide reliable support for various real-time applications. In the future, with the continuous evolution of technology, we believe that real-time task scheduling systems will continue to play a key role, driving innovation and development in embedded systems across various fields.

References

1. Kyrkou, C., Timotheou, S., Theocharides, T., Panayiotou, C., Polycarpou, M.M.: Optimizing multi-target detection in stochastic environments with active smart camera networks. In: Proceedings of the 11th International Conference on Distributed Smart Cameras (2017)
2. Debevec, P.E., Malik, J.: Recovering high dynamic range radiance maps from photographs. In: Proceedings of the 24th Annual Conference on Computer Graphics and Interactive Techniques (1997)
3. Reinhard, E., Stark, M.M., Shirley, P., Ferwerda, J.A.: Photographic tone reproduction for digital images. In: Proceedings of the 29th Annual Conference on Computer Graphics and Interactive Techniques (2002)
4. Carmona-Galán, R., Leñero-Bardallo, J.A., Fernández-Berni, J., Rodríguez-Vázquez, Á.: Pixel-wise parameter adaptation for single-exposure extension of the image dynamic range. In: Proceedings of the 10th International Conference on Distributed Smart Camera (2016)
5. Yang, M., Kriegman, D.J., Ahuja, N.: Detecting faces in images: a survey. IEEE Trans. Pattern Anal. Mach. Intell. **24**, 34–58 (2002)
6. Wu, Y., Ji, Q.: Facial landmark detection: a literature survey. Int. J. Comput. Vis. **127**, 115–142 (2018)
7. Alzu'bi, A., AlBalas, F., Al-Hadhrami, T., Younis, L.B., Bashayreh, A.: Masked face recognition using deep learning: a review. Electronics **10** (2021)
8. Liu, H., Zheng, N., Wang, Y., Li, J., Zhang, Z., Li, Y., Lan, J.: Development of a face recognition system and its intelligent lighting compensation method for dark-field application. IEEE Trans. Instrum. Meas. **70**, 1–16 (2021)
9. Gautam, K.S., Thangavel, S.K.: Video analytics-based intelligent surveillance system for smart buildings. Soft. Comput. **23**, 2813–2837 (2019)

Chapter 3
Universal Camera Jammer System

Abstract This chapter introduces a universal camera jammer system aimed at protecting privacy and security in the face of pervasive surveillance cameras. The system uses advanced algorithms and machine vision to detect and disrupt camera functionality. It combines hardware like a smart camera-steering system and laser interference to target and overexpose camera sensors, disabling them. The system merges technologies from machine vision, deep learning, and control engineering for quick camera localization and sustained interference. It includes a detailed description of its hardware components, such as the laser, smart camera, PTZ, and wheeled robot. Notable features are the enhanced YOLOv5 model for real-time camera detection and the Deep SORT algorithm for tracking multiple objects. The system assesses the camera's orientation and threat level to inform its interference tactics. It explains the self-feedback laser irradiation method, discussing the laser's impact on CMOS sensors and algorithms for accurate interference. In summary, the camera jammer offers a novel solution for personal privacy protection in surveillance-heavy settings, showcasing the potential of intelligent systems to counteract surveillance tech, balancing public safety with individual privacy.

Keywords Camera jammer · Privacy · Security · Laser interference · Camera localization

3.1 System Overview

With the advent of the digital age, surveillance equipment, especially universal cameras, has significantly improved public safety and personal convenience. However, it also raises widespread concerns about privacy infringement. These cameras (Fig. 3.1), including box cameras, bullet cameras, turret cameras, covert cameras, as well as wireless and network-powered cameras, greatly enrich the application fields of monitoring technology [1].

L. Wang et al., *Multi-spectral and Intelligent Sensing*,
SpringerBriefs in Computer Science, https://doi.org/10.1007/978-981-97-4550-0_3

Fig. 3.1 Cameras of different shapes

They typically use high-resolution RGB or RGB-IR sensors, capable of capturing clear images in well-lit environments and monitoring in low or no light conditions through infrared technology, thus achieving 24-h surveillance. However, the widespread popularity of universal cameras has gradually highlighted personal privacy and security issues. To address this challenge, we need an effective system to interfere with universal cameras, ensuring the protection of privacy and security.

Methods for interfering with universal cameras can include physical blocking, circuit damage, electromagnetic interference, and overexposure of the sensor with a bright light source. These methods have implementation difficulties, obvious damage traces, and the interference is easily detected. Manual handheld laser light source irradiation of surveillance cameras has some flaws, such as difficulty in timely detection of the opponent's camera (easy to miss), slow response speed, low irradiation precision control, difficulty in aiming, and scanning leakage. Therefore, we need a system capable of quickly and automatically locating, tracking, and precisely interfering with cameras in complex environments.

In this context, we conducted a study and designed a system using intelligent algorithms to detect and track universal cameras and interfere with them by precisely irradiating their lenses with lasers to cause overexposure (Fig. 3.2). This system has the capability to quickly locate and continuously interfere with different types of universal cameras, providing protection for personal privacy and security. The hardware of the system includes a smart camera-controlled steering system and an interference system (Table 3.1). Its core functions include detecting and tracking universal cameras within a preset area using machine vision technology, and judging and analyzing their monitoring range and threat level by determining the camera's posture. Finally, the camera is interfered with using laser irradiation technology, rendering it inoperable.

The design philosophy of this system is to provide a feasible solution for personal privacy and security under the premise of meeting legal and ethical requirements. With the application of intelligent algorithms, we can quickly identify and locate

Fig. 3.2 Universal camera jammer system architecture

Table 3.1 Description of the system hardware

Device	Device image	Parameter description
Laser		[Power] 30 mW [Voltage] DC 3.7 V [Size] 160 * 22 * 22 mm [Irradiation distance] 1000 m
Smart camera		[Power consumption] 5 W [Size] 75 * 50 * 37 mm [Image resolution] 1920 * 1080 [CPU] ARM 4 core 2.0G ARM Mali-G52 1Tops
PTZ		[Accuracy] ±0.1° [Size] 256 * 154 * 270 mm [Transmission mode] Worm gear and worm [Operating temperature] −25 to +55 °C [Speed] Level 0.02°–20°/S; pitch 0.02°–10°/S [Angle] Horizontal 0°–360° continuous rotation; pitch −60° to 20°
Wheeled robot		[Size] 400 * 300 * 200 mm [Drive mode] Motor [Operating temperature] −25 to +55 °C [Speed] 5 m/S [Load] 5 kg

universal cameras and precisely interfere with them using laser technology. This system design is not only suitable for public places and enterprises but also can be applied to personal homes and other fields. For those concerned about personal privacy and security, this system provides an effective protection method.

One of the key features of this system is the use of intelligent algorithms to detect and track universal cameras. With machine vision technology, the system can accurately identify the camera's location and track its movement in real time. This allows the system to quickly and automatically locate cameras in complex environments, regardless of their diverse types. Additionally, the system can analyze the camera's monitoring range and threat level, providing strong support for the formulation of interference strategies.

Figure 3.3 shows the workflow of a jammer system.

This system has been designed to provide an innovative solution for personal privacy and security, integrating technologies from multiple fields such as machine vision, deep learning, and control engineering, to achieve rapid positioning and continuous interference with universal cameras. It effectively addresses the privacy and security issues brought about by the widespread use of universal cameras. Whether in public places or private homes, this system offers reliable protection for people, ensuring that personal privacy is fully respected and protected. With the

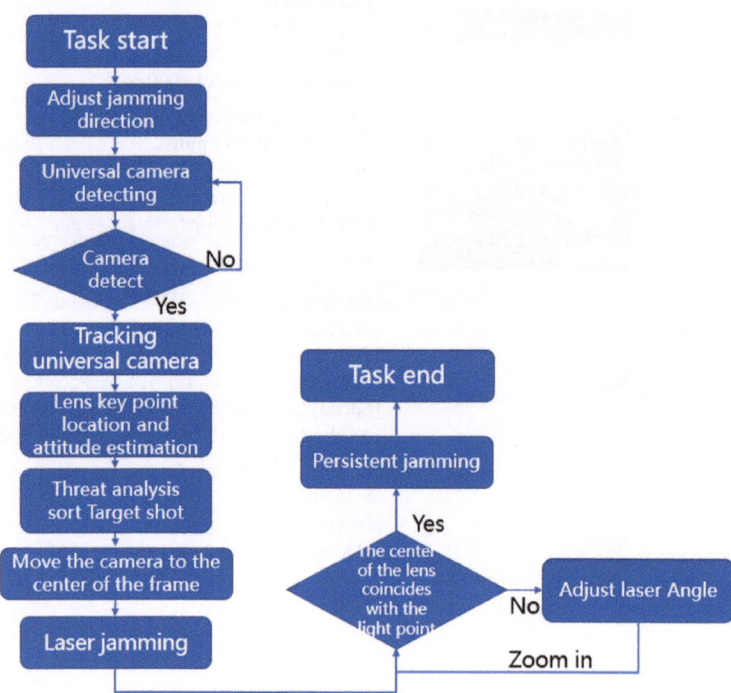

Fig. 3.3 Workflow of jammer system

continuous advancement of technology, we believe this system will play an increasingly important role in the future, contributing to the sustainable development of society.

3.2 Detection Cameras—Identification of Positioning and Attitude

3.2.1 Detection and Tracking Network Model

Universal cameras, a type of semiconductor imaging device, are known for their high sensitivity, resistance to strong light, minimal distortion, compact size, long lifespan, and vibration resistance. They are widely used in security systems for image acquisition and transmission. Universal cameras can convert image signals into digital signals and transmit them to remote monitoring centers via networks for real-time viewing by administrators. In addition to image acquisition, universal cameras also have functions such as image compression, storage, video analysis, and can perform operations like alarms, recording, and playback.

However, due to the diverse appearances of universal cameras and their installation in various directions and locations (Fig. 3.4), they are also easily obstructed by surrounding objects and affected by changes in lighting conditions. Therefore, long-distance detection of universal cameras poses a challenge.

Detection of cameras is the key first step in the system. Considering the complexity of universal camera types, we have collected more than 8,000 images of cameras of various shapes and annotated them, including the type of camera (bullet, dome, special shape), the external outline of the camera, the bounding box of the lens area, and landmark. To enhance the generalization ability of the detection model, we implemented data augmentation (Fig. 3.5) by superimposing camera regions onto various background images, using 30,000 training images and 4,500 validation images.

Deep Convolutional Neural Network (DCNN) is an important model in the field of deep learning, especially achieving significant results in computer vision [2]. Traditional object detection algorithms usually require manual feature design, whereas DCNNs can automatically learn features from images. Through multiple layers of convolution and pooling operations, DCNNs can extract hierarchical features from images, enabling the network to better understand image content. Additionally, DCNNs employ techniques like anchor mechanisms and non-maximum suppression to further improve the accuracy and robustness of object detection. To achieve real-time detection, we chose YOLOv5 as the base network. YOLOv5, an excellent version in the YOLO series, performs well in terms of speed and accuracy [3]. We built a deep learning model based on an improved YOLOv5 for real-time detection. Furthermore, we proposed improvements to the YOLOv5 network model. Despite YOLOv5's excellent performance in general object detection, to more precisely

Fig. 3.4 Cameras at different installation positions

locate cameras, we introduced an improved neural network (Fig. 3.6). This network adds extra convolutional layers on top of YOLOv5 to capture more detailed features related to cameras and introduced a landmark localization feature. Moreover, we introduced an attention mechanism, allowing the network to focus more on key areas in the image, further improving localization accuracy.

In the input size of 608 × 608, the improved neural network's performance in detecting cameras is as follows. By combining strategies such as sliding windows and Non-Maximum Suppression (NMS), detection across the full image of 1920 × 1080 is achieved. Table 3.2 describes the performance of the detection model.

3.2.2 Tracking Model

Deep SORT (Simple Online and Realtime Tracking) is an algorithm used for multi-object tracking in video streams [4]. It is an extension of the SORT (Simple Online and Realtime Tracking) algorithm, which uses the Kalman filter for object tracking. Building on the basis of universal camera detection, the Deep SORT algorithm can achieve continuous tracking of cameras through feature extraction and correspondence establishment between consecutive frames, ensuring the continuity of camera interference.

Fig. 3.5 Data augmentation

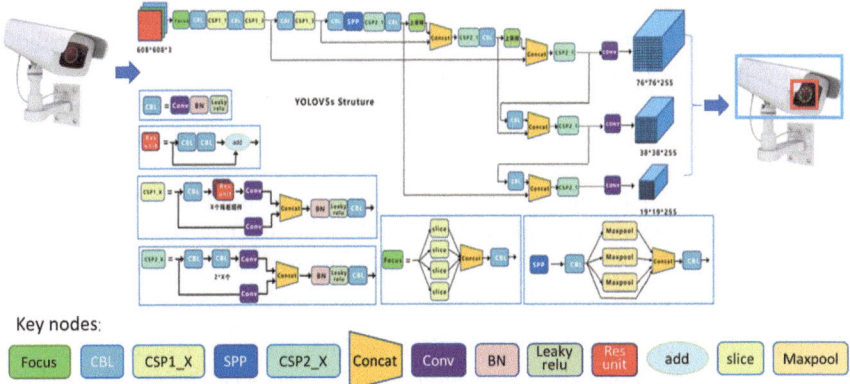

Fig. 3.6 Network structure

Table 3.2 Performance of camera detection model

Scales	mAP.5	mAP.5 small
S	0.932	0.927
M	0.9704	0.955
L	0.981	0.963

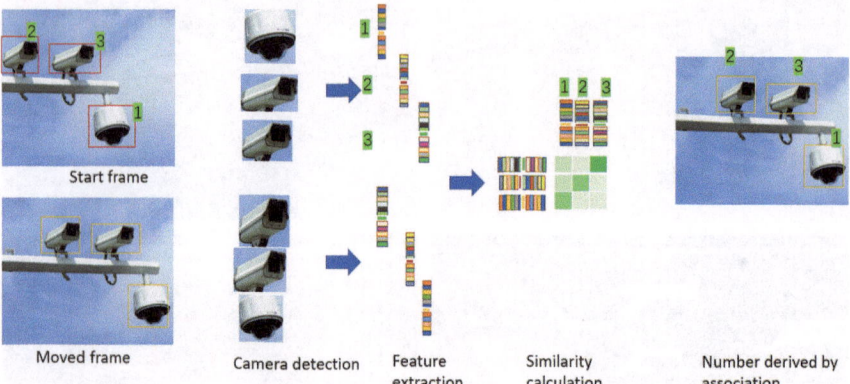

Start frame   Moved frame   Camera detection   Feature extraction   Similarity calculation   Number derived by association

Fig. 3.7 Camera tracking algorithm based on Deep SORT

Specifically, the main steps of the Deep SORT algorithm include (Fig. 3.7):

1. Object Detection: In the initial frame, use an improved YOLOv5 or other detection network models for universal camera detection. The detection model provides the camera's position and shape information.
2. Feature Extraction: Perform feature extraction on the camera regions detected in the initial frame. Typically, deep learning models like Convolutional Neural Networks (CNN) are used to extract image features, mapping the camera regions to the feature space.
3. Object Tracking: In subsequent consecutive frames, match the cameras with those in the previous frame through feature extraction and correspondence establishment. Various distance measurement methods (e.g., Euclidean distance, cosine similarity) can be used to measure the similarity between features and perform matching based on similarity.
4. State Update: Update the camera's position and state information based on the matching results. Filtering algorithms like Kalman filter or other filters can be used to predict and update the target's state, improving the tracking's stability and accuracy.

Similar to SORT, Deep SORT uses a Kalman filter for state prediction. The state of an object typically includes its position, velocity, and acceleration. The Kalman filter predicts the state of each object in the current frame based on its last known state, accounting for the object's motion dynamics.

Through these steps, the Deep SORT algorithm can achieve continuous tracking of multiple cameras. By extracting and matching features in consecutive frames, the Deep SORT algorithm maintains the consistency of camera numbers, ensuring continuous tracking of cameras. This method of continuous tracking can handle interference with cameras on the move, allowing the system to update and monitor the camera's position and state in real time.

3.2.3 Camera Pose Estimation and Threat Level Classification

After detecting and locating the camera, estimating the camera's pose and classifying its threat level are important steps in the universal camera jamming system. Camera pose estimation provides direction and angle information of the camera, further analyzing its monitoring range. Threat level classification, based on the results of pose estimation, evaluates the camera's potential threat level, providing decision-making basis for subsequent interference strategies. This section will detail the methods for camera pose estimation and threat level classification.

3.2.3.1 Coordinate System Establishment

To estimate the camera's pose, we first need to establish a coordinate system (Fig. 3.8). During the camera detection phase, we have already obtained the camera's image coordinates. By correlating with the actual scene, we can establish a mapping relationship between the camera image coordinate system and the world coordinate system. Taking the jamming system as the origin $O(0, 0, 0)$ of the world coordinate system, using a right-handed coordinate system, where the x-axis points to the right, the y-axis points upwards, and the camera's direction is the negative half-axis of the z-axis.

The opponent's universal camera also establishes a rectangular coordinate system based on the installation direction and axis direction, estimating its orientation through the camera's shape features, shadows, or known model parameters. This often involves complex image processing or machine learning technologies.

The field of view (FOV) of a generic camera, assuming the horizontal and vertical field of view angles (FOV) of each generic camera are known, represents the visual range that the camera can cover. Based on the position, orientation, and FOV angles of the generic camera, a conical area of visibility can be calculated. Any object within this conical area is considered to be visible by the generic camera. Figure 3.9 shows an example of a universal camera's field angle and monitoring distance.

Determining whether a smart camera is within the visible range of a generic camera can be simplified to determining whether the origin $O(0, 0, 0)$ is inside the generic camera's viewing cone.

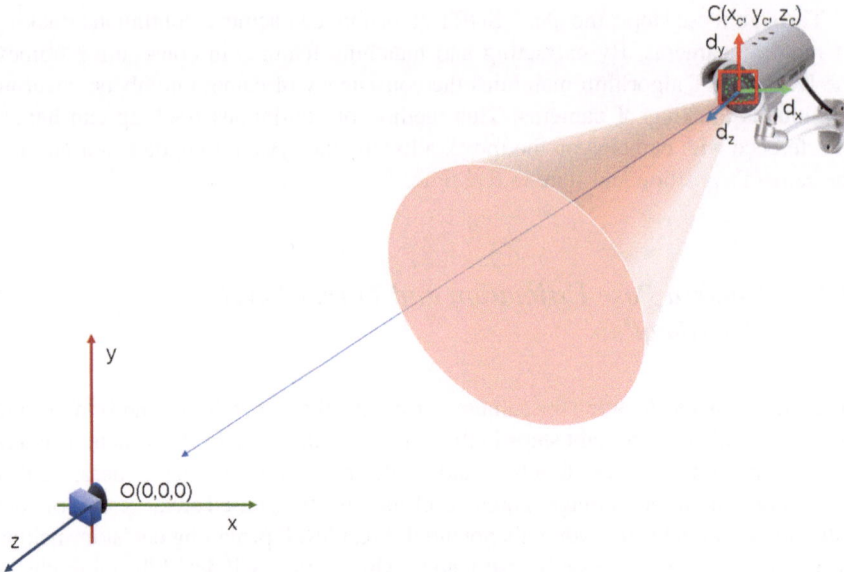

Fig. 3.8 Coordinate diagram of target camera and jamming camera

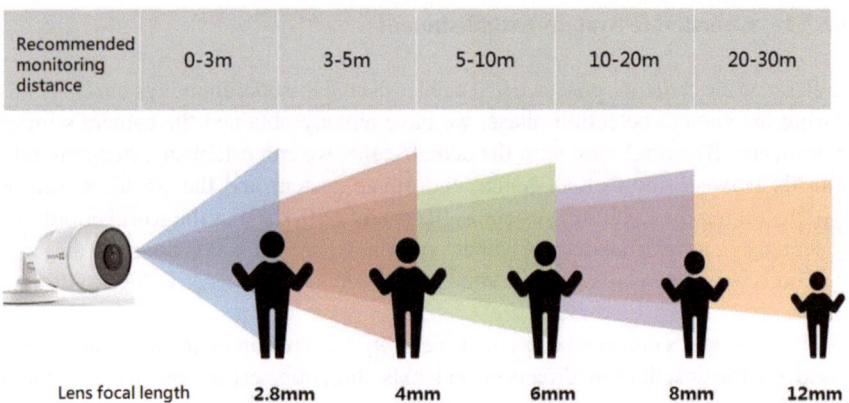

Recommended monitoring distance	0-3m	3-5m	5-10m	10-20m	20-30m
Lens focal length	2.8mm	4mm	6mm	8mm	12mm

Fig. 3.9 Field angle and monitoring distance of a universal camera

This can be judged through vector and angle calculations. First, calculate the angle between the vector pointing from the generic camera to the smart camera and the camera's orientation vector. If this angle is within half of the horizontal and vertical FOV, then the smart camera is within the visible range of the generic camera.

Assuming the position of the generic camera is $C(x_c, y_c, z_c)$, the orientation vector is $\vec{d} = (d_x, d_y, d_z)$, and the position of the smart camera is at the origin $O(0, 0, 0)$. The vector from the generic camera to the smart camera is $\vec{v} = (x_c, y_c, z_c)$. The angle θ between the two vectors can be calculated as follows:

$$\theta = \cos^{-1}\left(\frac{\vec{d} \cdot \vec{v}}{\|\vec{d}\| \|\vec{v}\|}\right)$$

where $\vec{d} \cdot \vec{v}$ represents the dot product of the vectors, and $\|\vec{d}\| \|\vec{v}\|$ are the magnitudes of the vectors, respectively. If θ is less than or equal to half of the camera's horizontal and vertical FOV, then the smart camera is within the visible range of the universal camera.

3.2.3.2 Threat Level Classification

Based on the results of camera pose estimation, we can further classify the threat level (Fig. 3.10). Threat level classification considers multiple factors, including the camera's monitoring range, resolution, and orientation.

By estimating the camera's monitoring range through its pose angle, we can generally say that the broader the monitoring range, the higher the potential threat level. Depending on the camera's model and parameters, we can set different thresholds to classify the monitoring range into different levels.

The camera's resolution is also an important factor in assessing the threat level. High-resolution cameras can capture clearer images, increasing the threat to privacy. We can determine the threat level by detecting and identifying the camera model to obtain its resolution information.

The camera's orientation also affects the threat level. If the camera directly faces sensitive areas or specific targets, its threat level is accordingly higher. Through the

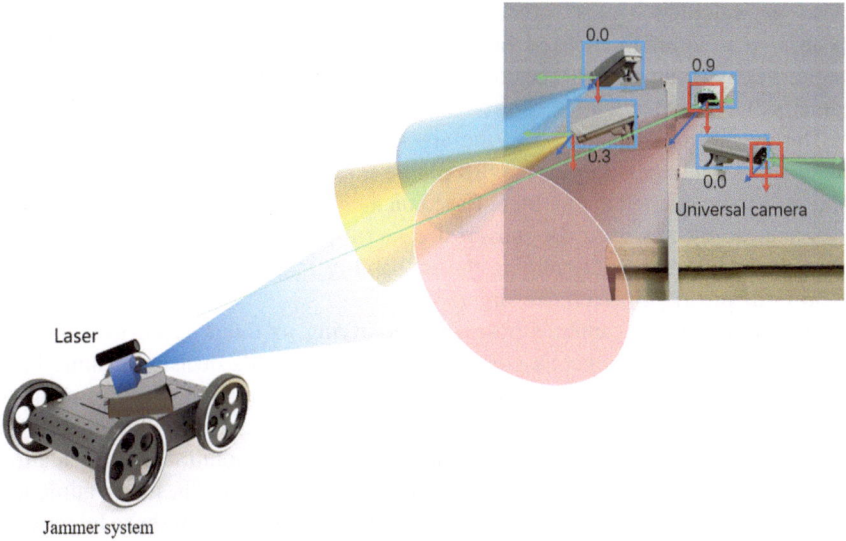

Fig. 3.10 Threat level judgment

results of pose estimation, we analyze the camera's orientation and adjust the threat level accordingly.

To calculate the probability of the universal camera observing the smart camera, we consider several key factors: the distance between the camera and the smart camera, the camera's field of view angles (horizontal and vertical), and the angle between the camera's orientation and the smart camera's direction. Calculating this probability requires integrating these factors, assuming a known or hypothesized mathematical relationship between the probability and these factors.

The probability is inversely proportional to the distance, meaning the farther the distance, the lower the probability of being observed.

The probability is directly proportional to the cosine of the angle, meaning the smaller the angle (closer to the camera's direct facing direction), the higher the probability of being observed.

Considering the above factors, we can use the following formula to calculate the camera's threat level:

$$p = k \times \cos(\theta) \times f(D)$$

where p is the probability of being observed, k is a proportionality constant, θ is the angle between the camera's orientation and the smart camera's direction, and $f(D)$ is a function of distance, representing the impact of distance on probability. We choose an exponential decay function for the distance function, for example: $f(D) = e^{-\lambda D}$. Integrating the above, the complete probability calculation formula becomes:

$$P = k \times \cos(\theta) \times e^{-\lambda D}$$

Here, the calculation of θ should consider the universal camera's field of view, ensuring θ is within the field of view range. If θ exceeds the camera's field of view angle, we can assume $P = 0$ (i.e., the smart camera is completely out of the universal camera's field of view).

During the implementation of the algorithm, parameters such as the common camera's size and field of view angles can be stored in an array, and the calculation process can be queried in real time. Based on the field of view of the smart camera's lens and the pixel size of the universal camera in the image, distance estimation is performed, thereby determining the probability ranking of being observed and identifying priority interference targets.

By estimating the camera's pose and classifying its threat level, we can more accurately assess the camera's potential threat and provide decision-making basis for subsequent interference strategies. Camera pose estimation employs methods for establishing coordinate systems and calculating pose angles, while threat level classification considers factors such as monitoring range, resolution, and orientation. The application of these technical means enhances the detection, localization, and analysis capabilities for cameras, effectively distinguishing interference targets.

3.3 Self-feedback Laser Irradiation Interference for Cameras

Once a universal camera is located, the system automatically interferes with it using laser irradiation. In the universal camera jamming system, self-feedback laser irradiation interference is a core component responsible for quickly and effectively interfering with detected and located universal cameras. To ensure the precision and efficiency of irradiation, we adopted a self-feedback control algorithm combined with a series of technical methods. Additionally, we expanded the beam to form a uniformly angled light spot covering the entire lens, effectively preventing the universal camera from escaping interference by adjusting its angle. The system can move with the mobile platform, continuously automatically interfering with all threatening universal cameras in view, thus fulfilling mission requirements. This section will detail the design, working principle, and implementation methods of self-feedback laser irradiation interference for cameras.

3.3.1 Laser Interference Principle

Universal cameras, depending on different field of view angles or lens focal lengths, have different detection angles and detection distances. Although flashlights and other light sources can cause saturation irradiation on cameras, laser irradiation is an effective method for non-contact interference with cameras from a distance without alarming or damaging them [5]. Laser interference irradiation on the CMOS sensor used by universal cameras can cause it to be overexposed and blinded (Fig. 3.11). Studies have shown that when the laser power exceeds the sensor's interference threshold, the irradiated area forms a nearly circular light spot that increases with laser power. This change is caused by the overall lifting of the Gaussian beam due to light intensity itself, rather than carrier diffusion. Therefore, when the incident laser beam is a Gaussian beam, its light intensity expression is:

$$I = I_0 e^{-\frac{2r^2}{\omega_0^2}}$$

To ensure the coverage effect of irradiation interference, it is necessary to consider the laser's power and adjust the laser lens's focusing degree to have a certain divergence effect, just covering the lens area of the distant universal camera.

Here, I represents the incident light intensity, r is the distance from the reference point to the center of the interference light spot, I_0 is the maximum light intensity at the center of the interference light spot, and ω_0 is the waist radius of the laser beam. Based on the relationship between the light intensity at the center of the interference light spot and the total incident laser power P, we can obtain:

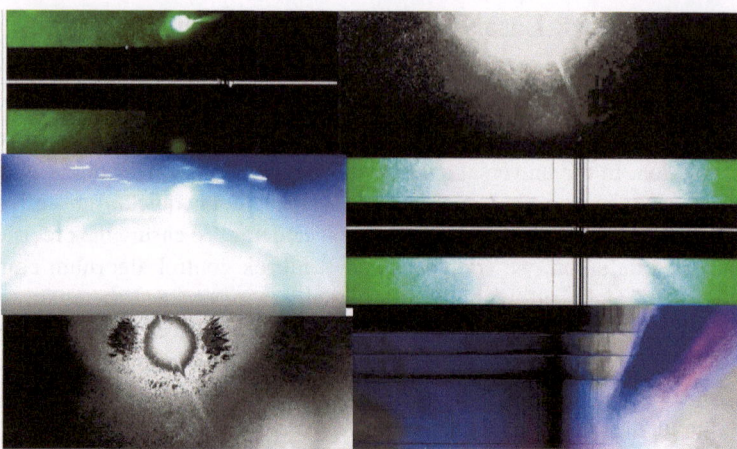

Fig. 3.11 The images after the camera is overexposed

$$P = \oint_0^\infty I_0 e^{\frac{-2r^2}{\omega_0^2}} \cdot 2\pi r \cdot dr = \frac{\omega_0^2}{2} \cdot \pi \cdot I_0$$

As the laser power increases, the maximum light intensity I_0 at the center of the interference light spot also increases, causing the overall lifting of the laser Gaussian beam due to light intensity itself. The distance r from the reference point to the center of the interference light spot increases, thus enlarging the interference light spot.

Due to the Gaussian distribution of the laser, where the intensity in the middle is much higher than the edges, black spots appear in the middle in all four images (Fig. 3.12). This indicates that due to the intensity being far above the critical value of saturation intensity, the middle part of the CMOS image sensor has been damaged. Further observation of the fourth image reveals that the halo transitions from dark to bright and then from bright to dark, forming a bright ring with the brightest middle ring. This indicates that the brightest middle part is the maximum light intensity that the sensor can accept. Exceeding this value causes certain damage to the sensor. Therefore, further increasing the light intensity does not increase but decreases the image brightness, entering a soft damage state. With further increased light intensity, the image output brightness further decreases, reaching a certain brightness level, the image appears entirely black, i.e., the inner ring edge. Continuing to increase will make no difference in the CMOS image sensor output, easily causing hard damage. Therefore, the phenomenon of a black center and the brightest middle ring in the image appears.

We can also select different laser wavelengths based on different task requirements, with some lasers being green or red, or choosing invisible infrared laser wavelengths (Fig. 3.13).

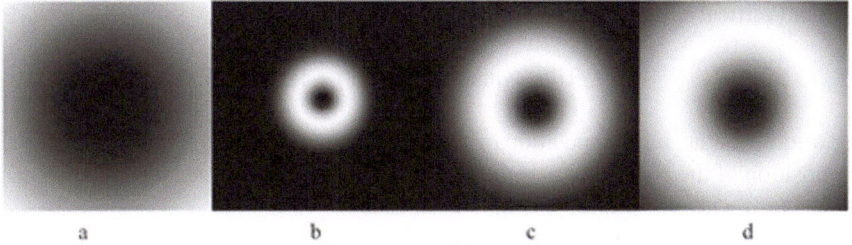

Fig. 3.12   Saturation effect of the CMOS image sensor under different light intensities

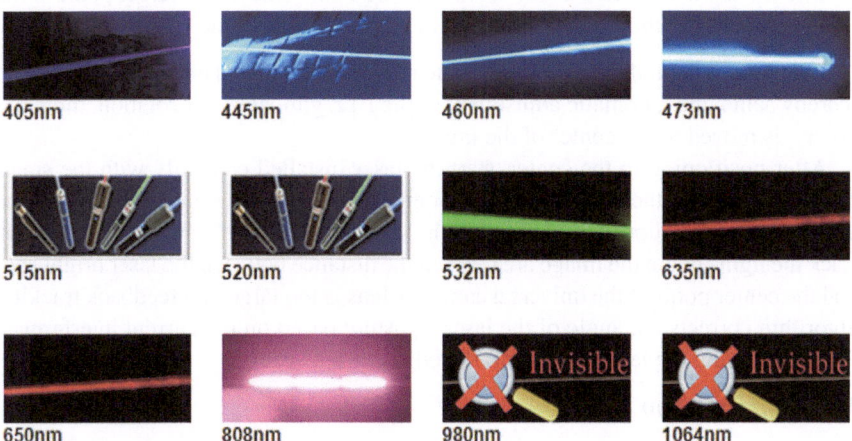

Fig. 3.13   Lasers of different wavelengths

3.3.2   Self-feedback Interference Algorithm

After deciding to interfere with a certain universal camera, the system controls the gimbal to rotate, aligning the smart camera with the universal camera to be interfered with, moving the universal camera to the center of the image. The calculation of the rotation angle can be based on the two-dimensional coordinate system, where the lens center point of the camera to the origin's x and y values can be calculated.

The formulas for calculating the necessary Pan/Tilt/Zoom parameters to rotate are as follows [6]:

- Pan (horizontal rotation) calculation formula:

$$\text{Pan} = a\tan^2(x - x_c, f_x) * 180/\pi$$

where x is the horizontal position of the target point in the image, x_c is the horizontal position of the center point of the image, and f_x is the camera's focal length.

- Tilt (vertical rotation) calculation formula:

$$\text{Tilt} = a\tan^2(y - y_c, f_y) * 180/\pi$$

where y is the vertical position of the target point in the image, y_c is the vertical position of the center point of the image, and f_y is the camera's focal length.
- Zoom (focal adjustment) calculation formula:

$$\text{Zoom} = k * (s/d)$$

where k is the camera's zoom coefficient, s s is the size of the target point in the image, and d is the distance between the target point and the camera.

Using these formulas, we can calculate the necessary Pan/Tilt/Zoom parameters, thereby achieving automatic adjustment of the PTZ gimbal. After rotation, the point (x_c, y_c) is moved to the center of the image.

After positioning to the center, start the laser installed coaxially with the smart camera to irradiate the opponent's lens for interference. At the same time, the irradiation image is continuously monitored by the smart camera. The system will detect and track the light spot in the image area. When the distance between the laser bright spot and the center point of the universal camera's lens is too large, the feedback tracking algorithm corrects the angle of the laser's posture based on the current interference image, ensuring the laser light is irradiated to the lens center.

Deviation Detection

The self-feedback control algorithm is the core of this system, detecting in real time the deviation between the laser spot irradiation position and the camera center, and adjusting the laser posture through fine-tuning to achieve automated closed-loop interference (Fig. 3.14).

After laser irradiation in the lens area of the universal camera, a light spot is formed, detected using rapid image processing methods. First, the image is converted from RGB to HSV space. Based on the chosen laser wavelength, the light spot color's HSV threshold range is determined, for example, a red or green light spot might be $(H_{red\_min}, H_{red\_max})$, with saturation and brightness also having corresponding ranges. Pixels within the threshold range in the HSV image are marked as 1 (white), others as 0 (black), and erosion and dilation operations can remove noise and fill in the light

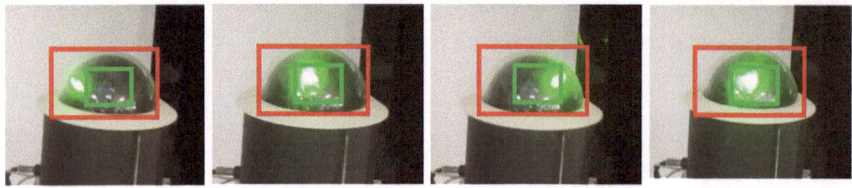

Fig. 3.14 Self-feedback interference

spot. Contour analysis and contour finding identify and mark continuous areas in the binary image. Ultimately, the center point of the light spot area is obtained.

Deviation detection is achieved through image processing technology. We first identify the positions of the laser spot and the camera in the image, then calculate the coordinate deviation between the center of the laser spot and the camera center. This deviation value serves as the input for the self-feedback control.

The formula for calculating coordinate deviation is as follows:

$$Dx = x_c - x_l$$

$$Dy = y_c - y_l$$

where (x_c, y_c) are the camera's center coordinates, and (x_l, y_l) are the laser spot's center coordinate.

Self-feedback Adjustment

Based on the deviation value, the self-feedback algorithm adjusts the gimbal's posture angle so that the laser spot emitted by the laser precisely hits the camera's lens. We use a PID (Proportional-Integral-Derivative) controller to implement self-feedback adjustment [7]. The output of the PID controller is the control signal for the gimbal, adjusting the gimbal's angle continuously until the laser spot precisely aligns with the camera center.

The PID controller formula is as follows:

$$u(t) = K_p e(t) + K_i \int_0^t e(\tau)d\tau + K_d \frac{d}{dt} e(t)$$

where $u(t)$ is the gimbal's control signal, $e(t)$ is the deviation value, K_p is the proportional coefficient, K_i is the integral coefficient, and K_d is the differential coefficient. By appropriately adjusting these three coefficients, we can achieve rapid and stable self-feedback control.

3.3.3 Interference System Effect

Through the design and implementation of self-feedback laser irradiation interference for cameras, we have successfully built an efficient and precise camera jamming system. The self-feedback control algorithm ensures the precision and stability of irradiation, while the beam expansion and light spot coverage technology enhance the interference effect. Continuous interference and automation make the system more practical and adaptable. The comprehensive application of these technical methods

Fig. 3.15 Interference system operating scenarios: disturbed image (left) and generic camera subjected to interference illumination (right)

provides strong support for protecting personal privacy and security. Figure 3.15 shows two application scenarios of the interference system.

Overall, this universal camera jamming system achieves rapid automation in locating universal cameras and continuous interference. Through intelligent control and closed-loop feedback, it can effectively and stably interfere with different types of universal cameras, providing excellent technical support for executing electro-optical countermeasure tasks. The system's design concept is flexible, and its modular components can be customized for a wider range of practical application scenarios.

References

1. Digital Cameras Market—Global Industry Analysis, Size, Share, Growth, Trends, and Forecast, 2021–2031 (n.d.). https://www.transparencymarketresearch.com/digital-cameras-market.html. Accessed 12 Dec 2021
2. Alzubaidi, L., Zhang, J., Humaidi, A.J., Al-dujaili, A., Duan, Y., Al-Shamma, O., Santamaría, J.I., Fadhel, M.A., Al-Amidie, M., Farhan, L.: Review of deep learning: concepts, CNN architectures, challenges, applications, future directions. J. Big Data **8** (2021)
3. Redmon, J., Divvala, S.K., Girshick, R.B., Farhadi, A.: You only look once: unified, real-time object detection. In: 2016 IEEE Conference on Computer Vision and Pattern Recognition (CVPR), pp. 779–788 (2015)
4. Parico, A.I., Ahamed, T.: Real time pear fruit detection and counting using YOLOv4 models and deep SORT. Sensors (Basel, Switzerland) **21** (2021)
5. Park, H., Choi, J., Choi, J.Y., Joo, K., Kim, N.: Investigation of the Hue–wavelength response of a CMOS RGB-based image sensor. Sensors (Basel, Switzerland) **22** (2022)
6. Sinha, S.N.: Pan-Tilt-Zoom (PTZ) camera. In: Ikeuchi, K. (ed.) Computer Vision. Springer, Cham (2020)
7. Zhong, J.: PID Controller Tuning: A Short Tutorial. Archived from the original (PDF) on 2015-04-21. Retrieved 2013-12-04 (Spring 2006)

Chapter 4
Biomedical Optical Sensors

Abstract In this chapter, we introduce the advantages of optical sensing technology and describe the typical biomedical optical sensor, photoplethysmography (PPG) device. Optical sensors have become an important tool in the field of modern medicine by virtue of their non-invasive, sensitive, real-time, and portable characteristics, and greatly promote the realization of remote physiological monitoring and telemedicine. In particular, we describe the integration of PPG technology in wearable devices, including principles, limitations, and future trends. Promising remote PPG technologies and their emerging applications are also introduced. Optical sensing technology is booming, and the accompanying challenges such as privacy and accuracy issues have attracted widespread attention.

Keywords Optical sensing · Biomedical sensor · Photoplethysmography · Remote physiological monitoring · Non-contact detection

4.1 Advantages of Optical Sensing Technologies

4.1.1 Introduction to Biomedical Optical Sensing

Biomedical optical sensors use light to collect information about biological tissues and have transformed the way medical diagnosis is performed. These sensors were originally designed to be simple, evolving from the foundations of photoplethysmography, which uses light to analyze changes in blood volume. Rapid advances in photonics and microelectronics have pushed these sensors to the forefront of medical innovation.

Today's optical sensors are complex devices capable of detecting multiple physiological indicators, providing high-resolution images of internal structures, and identifying biochemical markers in the body. This evolution has given rise to various types of sensors that are now indispensable tools in practices ranging from routine health monitoring to complex surgical procedures. As a result, biomedical optical sensors have become critical tools in the modern medical field, improving diagnostic capabilities and patient care.

4.1.2   The Evolution of Biomedical Optical Sensing

The development of biomedical optical sensing has gone from the early exploration of simple methods of using light to detect biological systems to the vigorous development today that is widely used in a variety of medical situations. Initially, these sensors primarily utilized the visible spectrum for non-invasive monitoring of cardiovascular health. As researchers explore other spectra, the use of technologies such as infrared spectroscopy makes it possible to monitor blood glucose levels without directly drawing blood [1].

The introduction of fiber optic technology in the 1960s greatly expanded the applications of optical sensors, allowing light to be transmitted flexibly and accurately to specific parts of the body [2]. Advances in laser technology have provided sensors with higher-resolution imaging capabilities and driven the development of minimally invasive surgical techniques.

Over the past few decades, the integration of computer-aided analysis and advanced imaging algorithms has significantly advanced the clinical use of biomedical optical sensors, allowing physicians to more accurately differentiate between healthy and diseased tissue. Today, biomedical optical sensing technology covers numerous technologies and instruments, such as endoscopes and wearable devices, enabling real-time monitoring of physiological changes [1, 2].

4.1.3   Non-invasiveness and Patient Comfort

The non-invasive nature of biomedical optical sensors has made a significant impact on patient comfort and healthcare delivery. These sensors do not require any incision or insertion into the body, thereby reducing the risk of infection and discomfort associated with invasive procedures. Non-invasiveness also leads to quicker recovery times and less stress for patients, as it eliminates the need for procedures like anesthesia or extended hospital stays.

Biomedical optical sensors enable continuous monitoring of vital signs and other health parameters without disrupting the patient's daily activities. For example, wearable devices that use optical sensing can track heart rate, oxygen saturation, or glucose levels with minimal intrusion [3]. This ease of use encourages better patient compliance and allows for more consistent data collection, which is crucial for accurate diagnosis and effective treatment planning.

4.1.4   Sensitivity and Specificity

In the field of biomedical optical sensing, sensitivity and specificity are key parameters that determine the performance and reliability of these devices. Sensitivity
refers to the ability of a sensor to correctly identify the presence of a physiological or
biochemical signal and to detect small changes or low levels of a target analyte with
high accuracy. High sensitivity is critical for early detection of disease, capturing
subtle abnormalities promptly and leading to better treatment outcomes.

Specificity is the sensor's ability to selectively identify the target signal among a
variety of similar signals or background noise. The high specificity ensures that the
sensor will not be mistakenly triggered by irrelevant substances, avoiding situations
that could lead to misdiagnosis. This is critical for accurately diagnosing a condition
and monitoring specific biomarkers or response to treatment.

Balancing sensitivity and specificity is an important challenge in the design of
biomedical optical sensors. Improving one parameter often sacrifices the other, so
optimization strategies must take into account its intended clinical application. For
example, in cancer diagnosis, high sensitivity may be prioritized to ensure that no
malignant cells are missed, whereas in chronic disease monitoring, specificity may
be more important to avoid false positives.

Advances in materials science, signal processing, and optical technologies have
enabled biomedical sensors to continue to improve in sensitivity and specificity, and
they have become indispensable tools for precision and personalized medicine.

4.1.5   Real-Time Analysis and Rapid Diagnostics

The ability to perform real-time analysis is another key advantage of biomedical optical sensors. This capability allows physicians to receive physiological and
biochemical data immediately without having to wait for laboratory analysis results.
Such instant feedback is critical for emergency medical situations and rapid disease
diagnosis, allowing immediate medical decisions and effective treatment.

Rapid diagnosis reduces patient wait times and uncertainty, thereby increasing
patient satisfaction and improving treatment outcomes. For example, pathogens or
disease markers can be detected in minutes using handheld optical devices [4], which
is much faster than traditional laboratory methods.

Advances in biomedical optical sensors have played an important role in
improving the accuracy and reliability of real-time analysis and rapid diagnosis.
As these sensors gain more and more attention in medical practice, they are expected
to become key tools in daily diagnosis and treatment and in the response to public
health emergencies in the future.

4.1.6 Portability and Preclinical Applications

The portability of biomedical optical sensors gives them unique advantages in preclinical applications [5]. Portability means these devices can be easily carried to the patient's location and are not limited to hospital or laboratory settings. This portable diagnostic tool can provide high-quality medical services in community clinics, homes, and remote areas, greatly improving the accessibility and efficiency of medical resources.

Preclinical applications allow doctors to test and evaluate patients immediately during their visit, allowing for faster disease screening and treatment. Portable devices are especially valuable for patients who live in remote areas or who don't have easy access to traditional medical facilities. In addition, these devices have shown their importance in responding to public health emergencies, disaster scenes, and military applications.

As technology continues to advance, biomedical optical sensors have improved in size, energy consumption, and cost, which has further contributed to their popularity in preclinical applications. Portable optical sensors are becoming increasingly smart, versatile and user-friendly, revolutionizing rapid on-site diagnosis and personalized medicine.

4.1.7 Multi-parameter Monitoring

Multi-parameter monitoring is an important direction in the development of biomedical optical sensor technology. This monitoring system is capable of tracking multiple biomarkers and physiological parameters simultaneously, providing physicians with a comprehensive view of health status. By collecting multiple types of data in one device, doctors can gain a more complete understanding of a patient's condition and how it interacts to affect overall health.

For example, a multi-parameter optical sensor might be able to measure blood oxygen saturation, heart rate, blood pressure and body temperature simultaneously [6]. This means a complete snapshot of vital signs can be obtained from a single testing session, optimizing the diagnostic process and making treatments more efficient and effective. This type of monitoring is especially valuable for patients with chronic conditions, as it supports ongoing health management and timely adjustments to treatment plans.

With the advancement of sensor and data analysis technology, the application of multi-parameter monitoring in clinical decision support systems has become increasingly mature. Smarter algorithms can not only process complex data sets but also provide predictive insights into changes in patient status, opening new pathways for early intervention and disease prevention.

4.1.8 Network Connectivity and Telemedicine

Network connectivity and telemedicine are an important direction in the development of today's biomedical optical sensing technology. They provide an innovative method that allows medical services to transcend the physical limitations of traditional hospitals and reach patients more conveniently and efficiently.

First, network connections enable optical sensors to transmit data to remote medical servers or cloud platforms in real time. Doctors and medical experts can instantly obtain patients' physiological information for remote diagnosis and treatment recommendations. This instant data flow provides patients with a faster medical response, especially in emergencies, allowing them to receive prompt attention and treatment from professionals.

Second, telemedicine facilitates the delivery of high-quality medical services in remote areas or places with fewer resources. Physiological parameters monitored through optical sensors can be sent to remote medical centers through the network, and doctors can evaluate the patient's health status based on these data and even guide local medical staff to carry out necessary interventions [7].

In addition, the combination of optical sensors and network connections promotes the development of personal health management. Patients can use portable or wearable optical sensing devices to monitor their health status in real time and upload the data to medical platforms through applications for tracking and analysis. This not only helps patients self-manage chronic diseases, but also provides doctors with continuous patient health data, which helps to better understand the changes in the patient's condition and formulate more accurate treatment plans.

In addition, with the development of Internet of Things technology, optical sensors can be connected with other medical equipment to form an intelligent medical network. Such a network can not only automatically collect and analyze data, but also implement more advanced functions, such as automatic drug distribution, health warnings, and long-term tracking of physiological parameters.

Overall, network connectivity and telemedicine have significantly improved the accessibility and efficiency of medical services, while also providing patients with a more personalized and convenient medical experience. With the continuous advancement of technology, optical sensors will be more widely used in the field of telemedicine in the future, playing a key role in improving the quality and efficiency of global medical services.

4.2 Photoplethysmography (PPG): An Example of the Rise of Wearable Devices

Photoplethysmography (PPG) is a classic non-invasive biosensing technology, which has been widely used in daily life [8] and clinical practice [9] with its advantages of low cost and convenience. PPG technology easily enables continuous physiological monitoring and is therefore integrated into popular wearable devices [8, 10], such as smartwatches, rings, and ear clips.

PPG wearable devices have good application prospects in both commercial and clinical applications, which can realize real-time physiological monitoring and timely health warnings. For example, PPG signals provide an analysis of the cardiopulmonary system and cardiovascular function. Basic physiological indexes, including heart rate [8], HRV, blood pressure [11], and respiration, and cardiovascular indications, including peripheral vascular disease [12] and arterial stiffness [13], are popular subjects in PPG research. Sleep monitoring [14] and sleep apnea detection [15] are another noteworthy practice for wearable PPG. In addition, PPG technology can monitor autonomic nervous system function, including assessing mental stress [16] and estimating fatigue [17].

4.2.1 Principle of PPG

PPG signals originate from changes in blood volume in microvascular tissue, which are mainly induced by pulsation. According to the relative position of the light source and the photodetector, contact PPG device has two modes: reflection and transmission. In the case of the fingertip, the light source and light receiver are on the same side of the skin as the former, while the latter expresses the signal of light penetrating out of the tissue. Green, red, and near-infrared light sources [18, 19] are the most common choices for this type of contact pulsation measurement.

The collected PPG waveform contains DC and AC components. The latter contains the pulsation information during each cardiac cycle. The non-pulsating DC component consists of relatively static blood absorption, including tissues, bones, and veins, and relatively low-frequency changes, such as respiration, sympathetic nervous system activity, and thermoregulation [9]. Therefore, PPG signals contain a wealth of physiological information.

As for reflect PPG or remote PPG (RPPG) system, the cardiac information of PPG signal is definitely attributable to the arterial pulsation, but the specific physiological mechanism remains unclear. Changes in blood volume caused by vasomotor of subcutaneous arterioles have conventionally been considered to be a direct factor in generating PPG signals. Moço's research shows that remote PPG systems can indeed detect arterial blood [20], supporting this hypothesis. Others have proposed different ideas about where and why light and blood volume interactions occur. Kamshilin et al. [21] proposed a tissue compression model, which posits that the remote PPG signal is mainly derived from the deformation of the intercapillary tissue caused by the transmural pressure of pulsating arteries. The aggregation process of red blood cells has also been considered as a possible source of PPG signaling [22].

4.2.2 Limitations and Challenges

Although wearable PPG devices have great application potential, they currently have several limitations. First, the contact measurement mode has inevitable drawbacks [10]. Usually, the probe of a sensor can only monitor the signal at a single spot. In order to realize multi-point detection and obtain more comprehensive pulse information, multi-sensor PPG systems have been proposed. However, the comfort of wearing is reduced and daily activities are partially constrained. Direct contact with the skin also means that this method is not suitable for patients with trauma, such as skin ulceration.

Second, PPG signals are highly susceptible to motion artifacts. The different pressures of contact measurements on the skin may also have an impact on PPG quality. For both contact and remote PPG, the quality issues are often similar and are discussed in detail in the next section on RPPG.

In addition, wavelength selection in the PPG configuration is critical to the pulse physiological information collected [23]. Considering the penetration depth of different wavelengths, the absorption characteristics of water and hemoglobin, green, red and near-infrared light sources are the most common choices, as mentioned above. However, PPG signals of different wavelengths have different physiological interpretation and stability, and multi-wavelength PPG system [24] is a possible choice to obtain more comprehensive pulse information.

4.2.3 Future Directions and Trends

Wearable PPG devices still have a way to go from research to widespread practice and application. But it is foreseeable that pervasive healthcare is one of the future application trends of PPG technology, especially with the assistance of AI and the support of big data [25]. Expanded physiological monitoring capabilities and data-driven health care interventions are both achievable expectations [24].

Improvements in hardware and software will also give PPG equipment more efficient performance. For example, leaps in optoelectronics and battery technology will reduce the cost and power of PPG equipment [25], advances in optical chip technology can give PPG a wider range of wavelength configurations, and breakthroughs in computing power will improve the operating efficiency and monitoring level of PPG [24].

Another crucial topic is basic research on the origin of PPG signaling, which essentially determines the authenticity and validity of physiological information. For example, Monte Carlo simulation is a feasible computational method to detect the origin of PPG [26]. The constructed phantoms are another practical experimental method to verify the different wavelength properties and physiological origin of PPG [25].

4.3 Remote Photoplethysmography (RPPG) Technology

The remote photoplethysmography (RPPG) technology is an extension of the basic contact PPG, which belongs to the reflective type. The RPPG signal is usually collected by a camera at a certain distance in the form of video. This remote measurement method is the perfect solution to the physical discomfort caused by contact PPG devices, and has a wide range of applications, such as surgery [27]. However, the light source selection of RPPG is not as flexible as that of contact PPG. Considering the general application environment, ambient light is the most common light condition in many studies [28].

4.3.1 Processing Framework

A key element of the RPPG system is the camera. Sun et al. [10] mentioned that the characteristics of the camera directly affect the quality of the recorded image and thus the physiological signal. Common video collectors are digital cameras [28] and webcams [29]. The latter is a cheaper and more available option. Poh et al. [29] achieved simultaneous collection of physiological signals from multiple people through video recorded by a basic webcam. Sun et al. [30] extracted RPPG signals from recorded online meetings to estimate stress.

After obtaining the video information, the scheme of processing the RPPG signal is followed [31] (Fig. 4.1). The first step is the selection and tracking of the region of interest (ROI), which is supported by many technologies in the field of computer vision. The VJ [32] method and the CSK [33] method are common choices for capturing ROI. Most studies of RPPG take faces as recording objects. However, factors such as involuntary head movement and facial expression change may disturb the signal acquisition results. Therefore, a reasonable ROI target is the basic requirement of RPPG. Given the flatness and smoothness of the acquisition area and the physiological basis of blood vessel distribution, the cheek and forehead [34] are the options of many researchers. Adaptive methods have also been proposed to filter out non-skin pixels [35].

Integrating pixel information in ROI is the next step to effectively reduce environmental and system instability, including light variation and system noise. The pixel averaging method was first proposed by Verkruysse et al. [28], that is, the pixel value of ROI is averaged, which is suitable for homogeneous lighting conditions. This approach is very easy to operate and is most common in research and practice. In the case of asymmetric light patterns, there are studies to separate pulse signals by the distribution of skin pixels [36].

The following step is to find the appropriate color channel, usually combining or fusing a pulse wave signal from multiple color channels, based on the RPPG principle and skin optical properties. Statistical separation technique and model-based analysis are two main methods of channel selection. The former is typified by

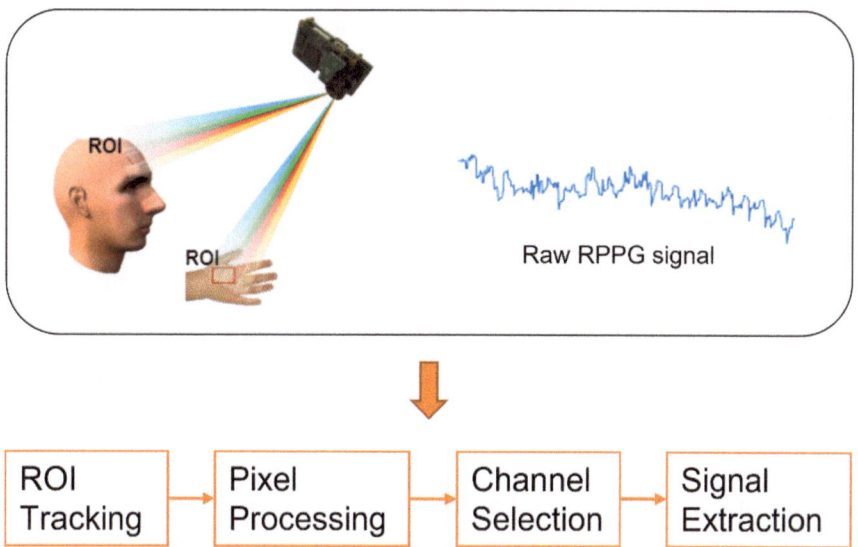

Fig. 4.1 Processing framework of RPPG system

blind source separation (BSS), a method for extracting physiological signals from multiple channels without prior knowledge. Classical ICA [37] and PCA [38] and their generalizations have been proposed to extract pulse information from RPPG signals.

The latter model-based approach is more faithful to physiological principles. It takes into account information about movement, environment, noise, and physiology. Generally speaking, the single-channel signal of RPPG is decomposed into pulsating component, static component, and disturbance component. Different reflection models may have different parameters and prior assumptions. de Haan et al. proposed chrominance model (CHROM) [39] and a blood volume change model (PBV) based on different wavelengths [40]. Wang et al. [41] in 2017 proposed a more general mathematical model using projective orthogonal planes (POS), capable of explaining BSS, CHROM, and PBV methods.

The last step is to filter the noise of the single channel signal and extract the hidden physiological information. The filtering technology is very diverse, among which bandpass filter [28], moving average filter [37], and wavelet transform are often applied to RPPG signal. Different digital filters have different noise reduction performance [42]. After noise reduction, physiological indicators of concern such as heart rate and respiratory rate will be extracted. This process can be implemented through interpolation, machine learning, and probabilistic algorithms.

The steps of the RPPG process are not necessarily in a fixed order. In the framework of deep learning, pioneering physiological information extraction methods may perform better [43]. The classic RPPG process may underestimate the full use of space–time and color space information. The underlying physiological information behind these complex signals has the opportunity to be explored through supervised learning and deep neural network models.

4.3.2   Challenges

RPPG system requires a complex combination of hardware and software, involving various fields. Although it enjoys great potential for practical application, RPPG technology still has many challenges at this stage [43]. There are many common considerations, not just for RPPG, but for basic contact PPG, or for remote visual physiological monitoring in general [44]. These factors are impediments to moving portable physiological monitoring technologies from the laboratory to clinical or practical applications.

There are many disturbing factors in the process of signal acquisition, including motion artifacts and illumination instability [43, 45]. These uncertainties interfere with RPPG model design and parameter selection. Many studies set up controlled conditions during the collection process, requiring subjects to maintain static or very limited head movement. But in some application scenarios, such as sleep and fitness monitoring, physical movement is unavoidable. Exploring more stable RPPG system design, tracking movement and compensating motion are two common strategies to deal with motion interference. The former includes the development of motion robust algorithms using spatial redundancy information [46], and the improvement of infrared lighting environments [47]. The latter involves designing specified motion postures during the experiment [48] and introducing additional sensors to capture motion information [49].

Lighting is another factor that is often difficult to control and predict in RPPG models. In some situations, such as driving, the quality of physiological signals can be seriously interfered with by various variations in lighting. Lin et al. [50] pointed out that light sources of different colors would affect the signal-to-noise ratio of RGB channels to varying degrees. To deal with the uncertainty and instability of illumination, the use of reference objects or the extraction of background illumination information is a possible solution. Li et al. [51] used the background region as a noise reference signal to compensate for lighting. Sun et al. [31] set different lighting conditions under experimental controlled conditions to reflect light information through smooth reference cards.

Another challenge is the fairness [43] and standardization of RPPG systems. Skin color is an important part of RPPG model building and signal acquisition. But most of the study subjects and data sets came from participants with light skin types. Differences in skin color can lead to biases in camera sensing, data set building, and physiological model design [43], while undermining the fairness and diversity of physiological measurements. Chari et al. [52] propose a physically-driven algorithm to improve the performance of HR estimation for facial videos with dark skin. In addition, Dasari et al. [52] collected facial video data from India and Sierra Leone and compared it with the ground truth signals. They demonstrate that the RPPG method is biased across different population groups and environmental conditions. The RPPG system lacks a recognized standard, which hinders its universality and practicality. The RPPG system lacks a recognized standard, which hinders its universality and practicality. The uncertainty and multi-choice of the composition of the RPPG system, including the style of the sensor, the setting of the light source, the specification of the video, the capability of calculation, the process of acquisition, the type of acquisition object, and the design of the experiment, make standardization difficult to achieve. This also undermines the reproducibility and accuracy of cross-sectional comparisons between studies.

The compression of PPG data and video is also a concern, especially in the data-driven era. In RPPG acquisition and processing operations, physiological information exists even at the pixel level. Therefore, the inappropriate method of data compression and decompression is likely to destroy the original content of pulse data, which will mislead the subsequent extraction of physiological indicators. Abdulkader and Qidwai [53] analyzed and compared several existing PPG compression techniques. Michal et al. [54] showed that the increase of compression rate will reduce the accuracy of HR estimation, and the application of color subsampling method can reduce the file size while retaining good heart rate estimation efficiency.

Privacy is a public topic, especially for human video capture. RPPG technology has the potential for non-invasive physiological monitoring at scale, but it also means the risk of personal information leakage. In the current situation where video physiological measurement technology is not popular and mature, any rash public, open data collection and technology application may cause serious impact. The tradeoff between the retention of physiological information and the protection of privacy is a serious issue. Gupta et al. [55] proposed a method of data perturbation that dramatically reduced the accuracy of facial recognition while maintaining the physiological content of RPPG. In order to prevent RPPG from being maliciously recorded in certain scenarios, such as online video conferencing, Sun et al. [56] proposed a method to protect physiological information. They modified the facial RPPG signal in the video based on convolutional neural networks.

4.3.3   Emerging Applications and Prospects

Despite the current challenges, the future development of RPPG technology and its ubiquitous application cannot be stopped. A variety of emerging applications are implemented and practiced on the basis of RPPG. Driver monitoring is an important application of RPPG to reduce accidents due to fatigue [57] or sudden illness. The study on the motion robustness of driving scene with the introduction of near infrared light source has achieved some results [58]. Sleep monitoring based on RPPG technology has also received attention. Polysomnography is the gold standard for sleep monitoring, involving multiple sensors but limiting the movement of the experimenter. van Gastel et al. [59] explored the feasibility of monitoring sleep disorders including sleep apnea through RPPG. They demonstrate the convenient diagnostic potential of cameras for sleep monitoring. Affective estimation is another popular application of RPPG, including mental load [60], depression [61], and stress [62, 63]. Multimodal datasets for stress recognition associated with physiological and psychological responses have been proposed [64].

Nowadays, digital health and telemedicine enjoy an increasingly important position in the field of health care, in which RPPG plays a considerable role. But there is still a gap from the laboratory to clinical practice [65]. This remote medical model has the advantages of low cost and low entry barrier, which is very suitable for home or community implementation. It is effective at all stages of disease prevention, diagnosis and treatment. At the same time, it significantly reduces the cost of data collection and promotes health analysis in the context of big data.

The standardization of the RPPG system also needs to be implemented in the future. Consistent understanding of RPPG is a powerful driver for the development of telemedicine, as well as for the theoretical optimization of physiological models. As discussed earlier with the challenges of RPPG, there are considerable difficulties in establishing standardization, including standard collection processes, software and hardware settings, and public privacy concerns. As technology and social awareness improve, these questions will be properly answered.

The consistency of datasets also belongs to the standardization category of RPPG system. The collection of large-scale, multimodal data sets is a time-consuming and labor-intensive task, and careful planning is important in advance. Excellent dataset saves a lot of time and effort for researchers, and provides significant help for the application practice of RPPG.

References

1. Tosi, D., Poeggel, S., Iordachita, I., Schena, E.: Fiber optic sensors for biomedical applications. In: Opto-mechanical Fiber Optic Sensors, pp. 301–333. Butterworth-Heinemann (2018)
2. Tuchin, V.V.: Tissue Optics. Society of Photo-Optical Instrumentation Engineers (SPIE), Bellingham, WA (2015)
3. Ballard, Z.S., Ozcan, A.: Wearable optical sensors. Mob. Health Sens. Anal. Methods Appl. 313–342 (2017)
4. Zarei, M.: Portable biosensing devices for point-of-care diagnostics: recent developments and applications. TrAC Trends Anal. Chem. **91**, 26–41 (2017)
5. Napp, J., Mathejczyk, J.E., Alves, F.: Optical imaging in vivo with a focus on paediatric disease: technical progress, current preclinical and clinical applications and future perspectives. Pediatr. Radiol. **41**, 161–175 (2011)
6. Pandian, P.S., Mohanavelu, K., Safeer, K.P., Kotresh, T.M., Shakunthala, D.T., Gopal, P., Padaki, V.C.: Smart vest: wearable multi-parameter remote physiological monitoring system. Med. Eng. Phys. **30**(4), 466–477 (2008)
7. Alenoghena, C.O., Ohize, H.O., Adejo, A.O., Onumanyi, A.J., Ohihoin, E.E., Balarabe, A.I., Alenoghena, B., et al.: Telemedicine: a survey of telecommunication technologies, developments, and challenges. J. Sens. Actuator Netw. **12**(2), 20 (2023)
8. Biswas, D., Simoes-Capela, N., Van Hoof, C., Van Helleputte, N.: Heart rate estimation from wrist-worn photoplethysmography: a review. IEEE Sens. J. **19**, 6560–6570 (2019)
9. Allen, J.: Photoplethysmography and its application in clinical physiological measurement. Physiol. Meas. **28**, 1–39 (2007)

10. Sun, Y., Thakor, N.: Photoplethysmography revisited: from contact to noncontact, from point to imaging. IEEE Trans. Biomed. Eng. **63**, 463–477 (2016)
11. Liu, J., Yan, B., Zhang, Y., Ding, X., Peng, S., Zhao, N.: Multi-wavelength photoplethysmography enabling continuous blood pressure measurement with compact wearable electronics. IEEE Trans. Biomed. Eng. **66**, 1514–1525 (2018)
12. Allen, J.: Photoplethysmography for the assessment of peripheral vascular disease. In: Allen, J., Kyriacou, P. (eds.) Photoplethysmography, pp. 189–235. Academic Press (2022)
13. Ferozoli, R., Karimpour, P., May, J., Kyriacou, P.: Arterial stiffness assessment using PPG feature extraction and significance testing in an in vitro cardiovascular system. Sci. Rep. **14**, 1–10 (2024)
14. Radha, M., Fonseca, P., Moreau, A., Ross, M., Cerny, A., Anderer, P., Long, X., Aarts, R.: A deep transfer learning approach for wearable sleep stage classification with photoplethysmography. npj Digit. Med. **4**, 1–11 (2021)
15. Hayano, J., Yamamoto, H., Nonaka, I., Komazawa, M., Itao, K., Ueda, N., Tanaka, H., Yuda, E.: Quantitative detection of sleep apnea with wearable watch device. PLoS ONE **15**, 1–14 (2020)
16. Charlton, P., Celka, P., Farukh, B., Chowienczyk, P., Alastruey, J.: Assessing mental stress from the photoplethysmogram: a numerical study. Physiol. Meas. **39**, 1–15 (2018)
17. Chen, Y.-X., Tseng, C.-K., Kuo, J.-T., Wang, C.-J., Chao, S.-H., Kau, L.-J., Hwang, Y.-S., Lin, C.-L: Fatigue estimation using peak features from PPG signals. Mathematics **11**, 3580 (2023)
18. Maeda, Y., Sekine, M., Tamura, T.: The advantages of wearable green reflected photoplethysmography. J. Med. Syst. **35**, 829–834 (2011)
19. Lee, J., Matsumura, K., Yamakoshi, K.-I., Rolfe, P., Tanaka, S., Yamakoshi, T.: Comparison between red, green and blue light reflection photoplethysmography for heart rate monitoring during motion. In: Annual International Conference of the IEEE Engineering in Medicine and Biology Society, pp. 1724–1727. IEEE Engineering in Medicine and Biology Society (2013)
20. Moco, A.V., Stuijk, S., De Haan, G.: New insights into the origin of remote PPG signals in visible light and infrared. Sci. Rep. **8**, 8501 (2018)
21. Kamshilin, A.A., Nippolainen, E., Sidorov, I.S., Vasilev, P.V., Erofeev, N.P., Podolian, N.P., Romashko, R.V.: A new look at the essence of the imaging photoplethysmography. Sci. Rep. **5**, 10494 (2015)
22. Batool, S., Nisar, M., Mangini, F., Frezza, F., Fazio, E.: Scattering of light from the systemic circulatory system. Diagnostics (Basel) **10** (2020)
23. Kim, K.B., Baek, H.J.: Photoplethysmography in wearable devices: a comprehensive review of technological advances, current challenges, and future directions. Electronics **12**, 2923 (2023)
24. Charlton, P.H., Allen, J., Bailón, R., Baker, S.B., Behar, J.A., Chen, F., Clifford, G.D., Clifton, D.A., Davies, H.J., Ding, C., Ding, X., Dunn, J.P., Elgendi, M., Ferdoushi, M., Franklin, D., Gil, E., Hassan, M.F., Hernesniemi, J.A., Hu, X., Ji, N., Khan, Y., Kontaxis, S., Korhonen, I., Kyriacou, P.A., Laguna, P., Lázaro, J., Lee, C., Levy, J., Li, Y., Liu, C., Liu, J., Lu, L., Mandic, D.P., Marozas, V., Mejía-Mejía, E., Mukkamala, R., Nitzan, M., Pereira, T., Poon, C.C., Ramella-Roman, J.C., Saarinen, H.J., Shandhi, M.M., Shin, H., Stansby, G., Tamura, T., Vehkaoja, A., Wang, W.K., Zhang, Y., Zhao, N., Zheng, D., Zhu, T.: The 2023 wearable photoplethysmography roadmap. Physiol. Meas. **44**, 1–89 (2023)
25. Kyriacou, P.A., May, J.M.: Photoplethysmography: new trends and future directions. In: Allen, J., Kyriacou, P. (eds.) Photoplethysmography. Academic Press (2022)
26. Chatterjee, S., Budidha, K., Kyriacou, P.A.: Investigating the origin of photoplethysmography using a multiwavelength Monte Carlo model. Physiol. Meas. **41**, 1–18 (2020)
27. Schraven, S.P., Kossack, B., Strüder, D., Jung, M., Skopnik, L., Gross, J., Hilsmann, A., Eisert, P., Mlynski, R., Wisotzky, E.L.: Continuous intraoperative perfusion monitoring of free microvascular anastomosed fasciocutaneous flaps using remote photoplethysmography. Sci. Rep. **13** (2023)
28. Verkruysse, W., Svaasand, L., Nelson, J.: Remote plethysmographic imaging using ambient light. Opt. Express **16**, 21434–21445 (2008)

29. Poh, M.-Z., Mcduff, D., Picard, R.: Non-contact, automated cardiac pulse measurements using video imaging and blind source separation. Opt. Express **18**, 10762–10774 (2010)
30. Sun, Z., Vedernikov, A., Kykyri, V.-L., Pohjola, M., Nokia, M.S., Li, X.: Estimating stress in online meetings by remote physiological signal and behavioral features. In: Adjunct Proceedings of the 2022 ACM International Joint Conference on Pervasive and Ubiquitous Computing and the 2022 ACM International Symposium on Wearable Computers (2022)
31. Sun, L., Wang, L., Shen, W., Liu, C., Bai, F.: Robust RPPG method based on reference signal envelope to improve wave morphology. Electronics **12**, 2813 (2023)
32. Viola, P., Jones, M.: Robust real-time face detection. Int. J. Comput. Vis. **57**, 137–154 (2004)
33. Henriques, J.F., Caseiro, R., Martins, P., Batista, J.P.: Exploiting the circulant structure of tracking-by-detection with kernels. In: European Conference on Computer Vision (2012)
34. Kwon, S., Kim, J., Lee, D., Park, K.S.: ROI analysis for remote photoplethysmography on facial video. In: 2015 37th Annual International Conference of the IEEE Engineering in Medicine and Biology Society (EMBC), pp. 4938–4941 (2015)
35. Fouad, R.M., Omer, O.A., Aly, M.H.: Optimizing remote photoplethysmography using adaptive skin segmentation for real-time heart rate monitoring. IEEE Access **7**, 76513–76528 (2019)
36. Wang, W., Stuijk, S., De Haan, G.: A novel algorithm for remote photoplethysmography: spatial subspace rotation. IEEE Trans. Biomed. Eng. **63**, 1974–1984 (2016)
37. Poh, M.-Z., Mcduff, D.J., Picard, R.W.: Advancements in noncontact, multiparameter physiological measurements using a webcam. IEEE Trans. Biomed. Eng. **58**, 7–11 (2011)
38. Lewandowska, M., Rumiński, J., Kocejko, T., Nowak, J.: Measuring pulse rate with a webcam—a non-contact method for evaluating cardiac activity. In: 2011 Federated Conference on Computer Science and Information Systems (FedCSIS), pp. 405–410 (2011)
39. Haan, G.D., Jeanne, V.: Robust pulse rate from chrominance-based rPPG. IEEE Trans. Biomed. Eng. **60**, 2878–2886 (2013)
40. Hann, D.G.G., Leest, V.A.A.: Improved motion robustness of remote-PPG by using the blood volume pulse signature. Physiol. Meas. **35**, 1913–1926 (2014)
41. Wang, W., Den Brinker, A.C., Stuijk, S., De Haan, G.: Algorithmic principles of remote PPG. IEEE Trans. Biomed. Eng. **64**, 1479–1491 (2017)
42. Guler, S., Golparvar, A.J., Ozturk, O., Doğan, H., Kaya Yapici, M.: Optimal digital filter selection for remote photoplethysmography (rPPG) signal conditioning. Biomed. Phys. Eng. Express **9** (2023)
43. Mcduff, D.J.: Camera measurement of physiological vital signs. ACM Comput. Surv. **55**, 1–40 (2021)
44. Huang, B., Hu, S., Liu, Z., Lin, C.-L., Su, J., Zhao, C., Wang, L., Wang, W.: Challenges and prospects of visual contactless physiological monitoring in clinical study. npj Digit. Med. **6** (2023)
45. Chen, X., Cheng, J., Song, R., Liu, Y., Ward, R.K., Wang, Z.: Video-based heart rate measurement: recent advances and future prospects. IEEE Trans. Instrum. Meas. **68**, 3600–3615 (2019)
46. Wang, W., Stuijk, S., Haan, G.D.: Exploiting spatial redundancy of image sensor for motion robust rPPG. IEEE Trans. Biomed. Eng. **62**, 415–425 (2015)
47. Van Gastel, M., Stuijk, S., De Haan, G.: Motion robust remote-PPG in infrared. IEEE Trans. Biomed. Eng. **62**, 1425–1433 (2015)
48. Chung, A.G., Wang, X.Y., Amelard, R., Scharfenberger, C., Leong, J., Kuliński, J., Wong, A., Clausi, D.A.: High-resolution motion-compensated imaging photoplethysmography for remote heart rate monitoring. In: Biomedical Optics and Imaging—Proceedings of SPIE, San Francisco, California, United States (2015)
49. Estepp, J.R., Blackford, E.B., Meier, C.M.: Recovering pulse rate during motion artifact with a multi-imager array for non-contact imaging photoplethysmography. In: 2014 IEEE International Conference on Systems, Man, and Cybernetics (SMC), pp. 1462–1469 (2014)
50. Lin, Y.-C., Lin, Y.-H.: A study of color illumination effect on the SNR of rPPG signals. In: 2017 39th Annual International Conference of the IEEE Engineering in Medicine and Biology Society (EMBC), pp. 4301–4304 (2017)

51. Li, X., Chen, J., Zhao, G., Pietikäinen, M.: Remote heart rate measurement from face videos under realistic situations. In: 2014 IEEE Conference on Computer Vision and Pattern Recognition, pp. 4264–4271 (2014)

52. Chari, P., Kabra, K., Karinca, D., Lahiri, S., Srivastava, D., Kulkarni, K., Chen, T., Cannesson, M., Jalilian, L., Kadambi, A.: Diverse R-PPG: Camera-Based Heart Rate Estimation for Diverse Subject Skin-Tones and Scenes. arXiv: Image and Video Processing (2020)

53. Abdulkader, S.S., Qidwai, U.A.: A review on PPG compression techniques and implementations. In: 2020 IEEE-EMBS Conference on Biomedical Engineering and Sciences (IECBES), pp. 511–516 (2020)

54. Rapczynski, M., Werner, P., Al-Hamadi, A.: Effects of video encoding on camera-based heart rate estimation. IEEE Trans. Biomed. Eng. **66**, 3360–3370 (2019)

55. Gupta, D., Etemad, A.: Privacy-preserving remote heart rate estimation from facial videos. In: 2023 IEEE International Conference on Systems, Man, and Cybernetics (SMC), pp. 706–712 (2023)

56. Sun, Z., Li, X.: Privacy-Phys: facial video-based physiological modification for privacy protection. IEEE Signal Process. Lett. **29**, 1507–1511 (2022)

57. Tsai, Y.-C., Lai, P.-W., Huang, P.-W., Lin, T.-M., Wu, B.-F.: Vision-based instant measurement system for driver fatigue monitoring. IEEE Access **8**, 67342–67353 (2020)

58. Nowara, E.M., Marks, T.K., Mansour, H., Veeraraghavan, A.: SparsePPG: towards driver monitoring using camera-based vital signs estimation in near-infrared. In: 2018 IEEE/CVF Conference on Computer Vision and Pattern Recognition Workshops (CVPRW), pp. 1353–135309 (2018)

59. van Gastel, M., Stuijk, S., Overeem, S., van Dijk, J.P., van Gilst, M.M., de Haan, G.: Camera-based vital signs monitoring during sleep—a proof of concept study. IEEE J. Biomed. Health Inform. **25**, 1409–1418 (2020)

60. Bousefsaf, F., Maaoui, C., Pruski, A.: Remote detection of mental workload changes using cardiac parameters assessed with a low-cost webcam. Comput. Biol. Med. **53**, 154–163 (2014)

61. Casado, C.Á., Cañellas, M.L., L'opez, M.B.: Depression recognition using remote photoplethysmography from facial videos. IEEE Trans. Affect. Comput. **14**, 3305–3316 (2022)

62. Nikolaiev, S., Telenyk, S., Tymoshenko, Y.: Non-contact video-based remote photoplethysmography for human stress detection. J. Autom. Mob. Robot. Intell. Syst. **14**, 63–73 (2019)

63. Morales-Fajardo, H.M., Arce, J.R., Gutiérrez-Cedeño, A., Viñas, J.C., Reyes-Lagos, J.J., Abarca-Castro, E.A., Ledesma-Ramírez, C.I., Gonzalez, A.V.: Towards a non-contact method for identifying stress using remote photoplethysmography in academic environments. Sensors (Basel, Switzerland) **22** (2022)

64. Meziatisabour, R., Benezeth, Y., de Oliveira, P., Chappé, J., Yang, F.: UBFC-Phys: a multimodal database for psychophysiological studies of social stress. IEEE Trans. Affect. Comput. **14**, 622–636 (2021)

65. Huang, B., Hu, S., Liu, Z., Lin, C., Su, J., Zhao, C., Wang, L., Wang, W.: Challenges and prospects of visual contactless physiological monitoring in clinical study. npj Digit. Med. **6**, 231 (2023)

Chapter 5
Healthcare—Perfect Realization of Intelligent Optical Sensors

Abstract In the modern medical background, the personalized, real-time and efficient healthcare has become an inevitable trend, in which intelligent optical sensors are indispensable thrust. The camera-based physiological monitoring method has liberated the inefficiency and inconvenience of traditional medical treatment, and has played a full role in today's medical field by virtue of its low cost and portability. With the development of technology, the concept of health has also been expanded. In this chapter, we describe non-contact routine physiological monitoring and a comprehensive approach to physiological signal analysis. We also introduce one of the future trends in health sensing, from passive blind stimulation to active real-time physiological feedback.

Keywords Intelligent sensor · Healthcare sensing · Non-contact · Physiological feedback · Remote monitoring

Intelligent optical biosensors have developed rapidly in recent decades, and occupy an increasingly important position in the field of healthcare [1]. Cameras [2] are the most common non-contact physiological monitors. The remote monitoring method liberates the contact discomfort and inconvenience from wearers of traditional biosensor, and has a wide range of application scenarios, including patient monitoring with open wounds and uncertain states of sleep and exercise. The advantages of low cost and convenience make camera sensors universal and valuable for mass production. The video signal acquisition mode of the smart camera can also realize the simultaneous physiological monitoring of multiple patients [3].

This chapter describes contactless monitoring of several typical routine physiological indicators, including respiratory movement, cardiovascular activity, blood pressure, blood oxygen saturation, and blood glucose. In addition, we present a comprehensive analysis of physiological signals and reveal possible future trends in health sensing—from passive blind stimulation to active real-time feedback.

5.1 Non-contact Monitoring of Routine Physiological Parameters

5.1.1 Respiratory Movement

Respiratory movement is one of the most concerned indicators in physiological moni-toring, which is directly related to physiological state of the human body. Small displacement detection of the region of interest (ROI) and optical flow techniques are common video-based methods to measure respiration, taking into account the motion characteristics in the respiratory signal.

Tan et al. [4] used differential information between frames to detect slight breath-induced changes in the chest and abdominal regions. But under unsatisfactory condi-tions, including clothes that lack high contrast and subjects or backgrounds that are not still, the study may produce erroneous measurements. Bartula et al. [5] proposed an algorithm based on vertical one-dimensional profile vectors to extract breathing-related motion and directionality to distinguish between inhalation and exhalation. This algorithm is more sensitive to tiny movements caused by breathing.

The optical flow method can describe the position change vector of the object between frames. It assumes that the brightness of pixels in areas of small local motion remains constant, matching the characteristics of breathing motion. Ganfure et al. [6] applied optical flow technology to continuous breathing monitoring and presented an algorithm with more efficient time complexity. Wang and Den Brinker [7] evaluated the performance of profile-based and optical flow algorithms in the context of magnetic resonance applications, and also discussed the advantages and limitations of various camera-based breath monitoring modes.

Other non-invasive camera-based respiratory monitoring methods have also been proposed. Janssen et al. [8] implemented an automatic ROI detection method for cameras based on the inherent characteristics of breathing, and extracted breathing signals based on motion decomposition. Van Gastel et al. [9] sought the best combi-nation of channels to suppress distortion, using the similarity of skin color changes caused by heartbeat and breathing. Low-frequency respiratory signals in this linear combination were then extracted.

5.1.2 Cardiovascular Activity

The cardiovascular system is the most critical physiological system in the body, transporting blood, nutrients and removing waste. Long-term monitoring of cardio-vascular activity is important, which can be achieved using remote camera-based systems. In the early years, Kranjec et al. [10] described and compared two methods of non-contact measurement of heart rate and heart rate variability, electrocardiogram and photoplethysmography (PPG). Al-Naji et al. [11] gave a comprehensive review of remote measurements of cardiopulmonary signals. In particular, they described the physiological background of the cardiovascular and respiratory systems and their physiological effects on the human body, including changes in skin color, pulsating arterial movements, and head and chest movements. These characteristic changes are the physiological mechanism of two common methods, ballistocardiography (BCG) [12] and PPG [13].

BCG characterizes the rigid movement of the body caused by the expulsion of blood from the heart [13]. When the BCG technique was applied to facial videos, the researchers focused on the regular small movements of the head that result from pumping blood through the carotid artery. This method is not limited to specific ROIs and can be used to detect heartbeat activity even when the skin is not visible. In 2013, Balakrishnan et al. [14] proposed a method for estimating heart rate from video of faces using BCG technology. They tracked the motion of feature points in the face region and carried out principal component analysis to extract the components of the time–frequency spectrum corresponding to the heartbeat. Haque et al. [15] used feature fusion technology to further weaken the feature loss caused by facial occlusion.

Remote photoplethysmography (RPPG) detects pulsating blood volume changes in the microcirculation under the skin. Verkruysse et al. [16] first realized heart rate extraction under ambient light using RPPG technology in 2008. Poh et al. [3] proposed a blind source separation method for face PPG signals and completed the work of simultaneous heart rate monitoring of multiple people. The introduction of additional band channels [17, 18] is another way to improve the extraction of cardiac activity signals based on video RPPG.

The jugular venous pulse (JVP) waveform contains information about heart health, but traditional tests require invasive catheter surgery. In recent years, non-invasive JVP monitoring has been implemented based on camera systems [19]. This could be complementary to BCG and PPG techniques and has the potential to bring more comprehensive information on cardiovascular activity.

5.1.3 Blood Pressure

Blood pressure refers to the pressure exerted by the flowing blood against the walls of blood vessels. Hypertension is the cause of many chronic diseases, and blood pressure monitoring is very important for disease prevention and treatment. Non-invasive methods of blood pressure measurement have been developed in recent years, including PPG and non-contact RPPG. Elgendi et al. [20] provided a comprehensive review of PPG techniques for the assessment of hypertension. They described methods of blood pressure measurement based on waveform analysis and wave propagation. The non-contact blood pressure monitoring method using camera system has made obvious progress, but there is still a long way to achieve the clinical measurement effect.

Pulse transit time (PTT) is a very important element in non-contact blood pressure measurement and describes the time regarding pulse propagation between two sites.

Shao et al. [21] proposed a camera-based real-time monitoring method for multiple physiological signals, including respiratory rate, expiratory flow rate, heart rate, and PTT, in 2013. In 2016, a study used high-speed cameras to extract PTT from faces and palms and verified the correlation between PTT and blood pressure in healthy subjects [22].

Waveform analysis is another common technique for measuring blood pressure. Schrumpf et al. [23] first trained the network on contact PPG data and then transferred it to RPPG-based blood pressure measurements. They also applied personalized techniques to retrain the neural network, reducing prediction errors.

5.1.4 Blood Oxygen Saturation

Blood oxygen measures the amount of oxygen in the blood and can be used for early detection of hypoxemia in the clinic. Arterial oxygen saturation (SaO2) represents the proportion of oxygen-saturated hemoglobin in arterial blood to total hemoglobin, and is typically estimated by percutaneous oxygen (SpO2) from pulse oximetry. This contact device uses two different wavelengths, generally 660 nm and 940 nm, corresponding to the maximum absorption wavelength of deoxygenated and oxygenated blood, respectively.

In 2005, Wieringa et al. [24] first demonstrated the possibility of measuring SpO2 remotely with a camera. They extracted the RPPG signal from the video at three different wavelengths (660, 810, and 940 nm). Humphreys et al. proposed a camera system in 2007 [25] with a dual-wavelength LED array (760 and 880 nm). They extracted physiological signals in a realistic light environment and presented a scheme for a remote pulse oximeter. Shao et al. [26] proposed a pair of wavelengths (611 and 880 nm) different from contact SpO2 monitoring based on the best signal-to-noise ratio, which better matches the measurement mode of remote video. Dual- or multi-wavelength signals are generally the basis for camera-based SpO2 remote measurements. There may be different optimal dual wavelength choices in different experimental settings and environments.

Kong et al. [27] used two cameras to simultaneously capture two different wavelengths (660 and 520 nm) of RPPG signals through narrowband filters. They achieved a non-contact measurement of blood oxygen saturation in the regular lighting environment. Tarassenko et al. [28] used red and blue channels captured by cameras to track changes in blood oxygen saturation during obstructive sleep apnea, using ambient light and an autoregressive model. But their approach required pulse oximeters to calibrate remote measurements of blood oxygen, limiting the application scenarios. Remote blood oxygen measurement systems with multiple wavelengths are easily disturbed by motion. Van Gastel et al. [29] introduced a new principle based on prior information about the relative pulse amplitudes of different wavelengths. They demonstrated the robustness of remote measurements of SpO2 during significant motion under this principle.

5.1.5 Blood Glucose

Diabetes is a serious metabolic disease with a significant increase in prevalence worldwide. Monitoring and controlling glucose levels in the blood is especially critical. Typical blood glucose monitoring techniques are invasive or minimally invasive. However, video-based non-invasive blood glucose monitoring devices have not been reliable until now. Wang et al. [30] demonstrated that using regular cameras to estimate blood glucose levels is not feasible, through simulation experiments in the visible to near-infrared wavelength range. They argued that the optical effect of glucose changes is much smaller than that of other physiological factors, including skin temperature, blood oxygen saturation, and water concentration. Near-infrared (NIR) monitoring techniques are challenging because the weak spectrum of glucose overlaps with the spectrum of other active chromophores [31].

Reliable non-invasive blood glucose monitoring systems are still being explored. In 2015, Yadav et al. [31] introduced monitoring technology based on NIR spectroscopy, including its limitations and challenges. Villena Gonzales et al. [32] provided a comprehensive review of techniques, devices, and algorithms for non-invasive and minimally invasive glucose monitoring. Alsunaidi et al. [33] reviewed the characteristics, advantages, and limitations of non-invasive optical glucose monitoring technologies from 2015 to 2021, including near-infrared, mid-infrared, far-infrared, and Raman spectroscopy.

5.2 Comprehensive Analysis of Physiological Signals

PPG and RPPG devices are playing an increasingly important role in healthcare with their low cost and convenience. In many cases, pulse signals are collected to target specific physiological indicators, such as heart rate. However, the health information contained in PPG and RPPG is actually very rich, and the integrity of the information may be destroyed in the process of pursuing specific physiological variables. A comprehensive analysis of physiological signals may restore authentic health information.

PPG signals express a lot of information about the pulse, including waveform, amplitude, energy, and envelope. The envelope is a global physiological feature that provides possible insights for the holistic analysis of pulse waves. As a remote non-contact technology, RPPG is more susceptible to disturbance than contact PPG. We proposed a method based on reference signal envelope to obtain better RPPG channel signals [34].

Compared with contact PPG signals, RPPG signals contain more noise and have poorer waveform quality, resulting in the derived physiological indicators may be inaccurate. Based on more stable contact PPG signals, it is possible to restore RPPG signals to high quality. Kim et al. [35] selected corresponding feature points of contact PPG and remote PPG signals, recovered RPPG signals and improved waveforms by using support vector regression and deep learning models. Haugg et al. [36] mentioned that RPPG signals were often compared to target physiological markers rather than to baseline PPG signals. Low-quality RPPG signals are sufficient to measure heart rate, but capturing more complex health information requires higher quality RPPG signals. They evaluated and compared the results of several common non-deep learning RPPG methods with reference fingertip PPG signals for similarity. These non-deep learning methods typically obtain a linear combination of the red (R), green (G), and blue (B) channels based on physiological principles, such as the chrominance-based (CHROM) [37] and plane-orthogonal-to-skin (POS) [38] methods.

We linearly decomposed the RPPG signal into pulsation, constant, and disturbance components, applying the physiological model of RPPG proposed by Wang et al. [38]. The optimal linear combination of R, G, and B channels was obtained by referring to the envelope shape of the PPG signal on the fingertip, which improved the waveform shape in an overall sense. Specifically, we obtained a set of coefficients, such that the envelope of the linear combination signal of the three channels had the greatest correlation coefficient with the envelope of the reference PPG signal. We designed four lighting conditions, corresponding to different environmental stability, and tested the approximation of the results to the baseline PPG curve based on the envelope-based (EB) method, CHROM, and POS algorithms.

In the four experimental environments, the EB method was robust and statistically better than CHROM and POS methods in envelope approximation. In addition, EB method improved the envelope morphology of the original green channel from the visual level. In these experiments, EB method contributed to morphological optimization in four aspects: correcting signal trend and direction, adjusting fluctuation scale, reducing noise, and enhancing single wave detail.

The improved RPPG method based on envelope morphology restores the global scale information of RPPG signal with physiological principle. On the basis of high-quality RPPG signals, more complex health information, such as emotions, may be detected remotely.

5.3 Future Trends in Healthcare Sensing

Traditional wearable devices or physiological sensing technologies focus on detecting specific health indicators in the human body. With the rise of intelligent sensing, this passive data acquisition mode has a tendency to transform into an active feedback form. As early as 1975, clinical biofeedback techniques were described with three basic principles [39]: the dynamic correlation of physiological states with emotional states, the individual's voluntary control over visible physiological states, and the facilitation of a deeply relaxed meditative state to establish such control. Today, biofeedback has a broader scope and can assist in the treatment or relief of many physical and mental problems, including stress and anxiety [40].

Nogueira et al. [41] described a growing interest in improving the quality of personal life, in the modern context of a "health culture". In 2018, they assessed the current state of biofeedback, reviewing existing medical grade and consumer grade hardware. Biofeedback devices are usually medical grade, expensive and inconvenient. Cheaper and more flexible smart sensing devices, such as smartwatches, could be the new solution for biofeedback. There is also a tendency for biofeedback to move from patient-only promotion to the general population, which is consistent with modern medical contexts and health concepts. They also analyzed consumer-grade physiological monitoring devices through several dimensions: price, available sensors, derived variables, existing API, software suites, operating system compatibility, communication protocols, and battery life. These considerations could affect the market value of small smart sensing technologies, the efficiency of physiological feedback, and their compatibility with medical grade devices.

The indirect or direct combination of biofeedback technology and consumer portable devices has been gradually explored in recent years. Can et al. [42] proposed a personalized stress monitoring and relaxation advice system utilizing consumer-grade smart bands. Chronic or acute stress has been linked to many diseases [43, 44], and regulating and managing stress is an important technique for maintaining physical and mental health. Traditional yoga [45] and mindfulness [46] are both beneficial ways to relieve stress, but are not usually suitable for office environments or daily routines. Can et al. [42] monitored stress based on a machine learning model of physiological signals, and then analyzed contextual information by electrodermal activity signals and provided flexible recommendations for conventional or mobile stress reduction. Rudics et al. [47] adopted a simpler biofeedback scheme, HRV-based respiratory regulation. HRV training achieved a reduction in stress and anxiety levels [40] by synchronizing breathing and heart rate, similar to the relaxation effects of methods such as mindfulness [48]. The combination of PPG technology and breathing training techniques is a potentially effective solution for health care smart sensing implementation to passively collect physiological signals and then actively provide real-time feedback.

The intelligent biosensor system for signal acquisition and real-time feedback of psychological stress has gradually received attention. There are many other psychological or physical states of concern, as well as alternative regulatory techniques or interventions. For example, different studies have illustrated the possibility of video-based RPPG technology to detect fatigue [49] and the effect of heart rate biofeedback strategies on improving exercise behavior under mental fatigue [50]. The use of a single smart sensor to achieve the detection and reduction of fatigue is worth exploring, which works for other mental states, including depression [51]. HRV-based biofeedback is also effective for physiological states, such as improving recovery after exercise [52]. In addition to respiratory regulation, other simple and effective biofeedback techniques have been developed. Kantan et al. [53] proposed a technical framework for using musical biofeedback to promote stroke recovery. Preliminary trials in subacute stroke patients have demonstrated the clinical significance of music-guided exercise in training.

An important consideration is that the detection of mental states from physiological signals is still under exploration and has not yet been fully validated in medical practice. Designing efficient, precise, personalized biofeedback strategies is also difficult. The combination of the two, that is, the realization of passive physiological signal acquisition and active real-time feedback through smart sensors, is challenging but valuable and meaningful.

References

1. Zhang, X., Hu, M., Zhang, Y., Zhai, G., Zhang, X.-P.: Recent progress of optical imaging approaches for noncontact physiological signal measurement: a review. Adv. Intell. Syst. **5**, 2200345 (2023)
2. Mcduff, D.J.: Camera measurement of physiological vital signs. ACM Comput. Surv. **55**, 1–40 (2021)
3. Poh, M.-Z., Mcduff, D., Picard, R.: Non-contact, automated cardiac pulse measurements using video imaging and blind source separation. Opt. Express **18**, 10762–10774 (2010)
4. Tan, K.S., Saatchi, R., Elphick, H.E., Burke, D.: Real-time vision based respiration monitoring system. In: 2010 7th International Symposium on Communication Systems, Networks & Digital Signal Processing (CSNDSP 2010), pp. 770–774 (2010)
5. Bartula, M., Tigges, T., Muehlsteff, J.: Camera-based system for contactless monitoring of respiration. In: 2013 35th Annual International Conference of the IEEE Engineering in Medicine and Biology Society (EMBC), pp. 2672–2675 (2013)
6. Ganfure, G.O.: Using video stream for continuous monitoring of breathing rate for general setting. Signal Image Video Process. 1–9 (2019)
7. Wang, W., Den Brinker, A.C.: Chapter 4—Camera-based respiration monitoring: motion and PPG-based measurement. In: Wang, W., Wang, X. (eds.) Contactless Vital Signs Monitoring. Academic Press (2022)
8. Janssen, R.G.J., Wang, W., Moco, A.V., De Haan, G.: Video-based respiration monitoring with automatic region of interest detection. Physiol. Meas. **37**, 100–114 (2016)
9. Van Gastel, M., Stuijk, S., De Haan, G.: Robust respiration detection from remote photoplethysmography. Biomed. Opt. Express **7**(12), 4941–4957 (2016)
10. Kranjec, J., Beguš, S., Geršak, G., Drnovšek, J.: Non-contact heart rate and heart rate variability measurements: a review. Biomed. Signal Process. Control **13**, 102–112 (2014)
11. Al-Naji, A., Gibson, K., Lee, S.H., Chahl, J.: Monitoring of cardiorespiratory signal: principles of remote measurements and review of methods. IEEE Access **5**, 15776–15790 (2017)
12. Inan, O.T., Migeotte, P.F., Park, K.S., Etemadi, M., Tavakolian, K., Casanella, R., Zanetti, J., Tank, J., Funtova, I., Prisk, G.K., Rienzo, M.D.: Ballistocardiography and seismocardiography: a review of recent advances. IEEE J. Biomed. Health Inform. **19**, 1414–1427 (2015)
13. Hassan, M.A., Malik, A.S., Fofi, D., Saad, N., Karasfi, B., Ali, Y.S., Meriaudeau, F.: Heart rate estimation using facial video: a review. Biomed. Signal Process. Control **38**, 346–360 (2017)
14. Balakrishnan, G., Durand, F., Guttag, J.: Detecting pulse from head motions in video. In: 2013 IEEE Conference on Computer Vision and Pattern Recognition, pp. 23–28 (2013)
15. Haque, M.A., Irani, R., Nasrollahi, K., Moeslund, T.B.: Heartbeat rate measurement from facial video. IEEE Intell. Syst. **31**, 40–48 (2016)
16. Verkruysse, W., Svaasand, L.O., Nelson, J.S.: Remote plethysmographic imaging using ambient light. Opt. Express **16**(26), 21434–21445 (2008)
17. Mcduff, D.J., Gontarek, S., Picard, R.W.: Improvements in remote cardiopulmonary measurement using a five band digital camera. IEEE Trans. Biomed. Eng. **61**, 2593–2601 (2014)
18. Xu, M., Zeng, G., Song, Y., Cao, Y., Liu, Z., He, X.: Ivrr-PPG: an illumination variation robust remote-PPG algorithm for monitoring heart rate of drivers. IEEE Trans. Instrum. Meas. **72**, 1–10 (2023)

19. Amelard, R., Hughson, R.L., Greaves, D.K., Pfisterer, K.J., Leung, J., Clausi, D.A., Wong, A.: Non-contact hemodynamic imaging reveals the jugular venous pulse waveform. Sci. Rep. **7** (2016)

20. Elgendi, M., Fletcher, R.R., Liang, Y., Howard, N., Lovell, N.H., Abbott, D., Lim, K., Ward, R.K.: The use of photoplethysmography for assessing hypertension. NPJ Digit. Med. **2** (2019)

21. Shao, D., Yang, Y., Liu, C., Tsow, F., Yu, H., Tao, N.: Noncontact monitoring breathing pattern, exhalation flow rate and pulse transit time. IEEE Trans. Biomed. Eng. **61**, 2760–2767 (2014)

22. Jeong, I.C., Finkelstein, J.: Introducing contactless blood pressure assessment using a high speed video camera. J. Med. Syst. **40**, 1–10 (2016)

23. Schrumpf, F., Frenzel, P., Aust, C., Osterhoff, G., Fuchs, M.: Assessment of non-invasive blood pressure prediction from PPG and rPPG signals using deep learning. Sensors **21**, 6022 (2021)

24. Wieringa, F.P., Mastik, F., Steen, A.F.W.V.D.: Contactless multiple wavelength photoplethysmographic imaging: a first step toward "SpO2 camera" technology. Ann. Biomed. Eng. **33**, 1034–1041 (2005)

25. Humphreys, K., Ward, T., Markham, C.: Noncontact simultaneous dual wavelength photoplethysmography: a further step toward noncontact pulse oximetry. Rev. Sci. Instrum. **78** (2007)

26. Shao, D., Liu, C., Tsow, F., Yang, Y., Du, Z., Iriya, R., Yu, H., Tao, N.: Noncontact monitoring of blood oxygen saturation using camera and dual-wavelength imaging system. IEEE Trans. Biomed. Eng. **63**, 1091–1098 (2016)

27. Kong, L., Zhao, Y., Dong, L., Jian, Y., Jin, X., Li, B., Feng, Y., Liu, M., Liu, X., Wu, H.-C.: Non-contact detection of oxygen saturation based on visible light imaging device using ambient light. Opt. Express **21**(15), 17464–17471 (2013)

28. Tarassenko, L., Villarroel, M., Guazzi, A., Jorge, J., Clifton, D.A., Pugh, C.W.: Non-contact video-based vital sign monitoring using ambient light and auto-regressive models. Physiol. Meas. **35**, 807–831 (2014)

29. Van Gastel, M., Stuijk, S., De Haan, G.: New principle for measuring arterial blood oxygenation, enabling motion-robust remote monitoring. Sci. Rep. **6** (2016)

30. Wang, Y., Wang, W., Gastel, M.V., Haan, G.D.: Modeling on the feasibility of camera-based blood glucose measurement. In: 2019 IEEE/CVF International Conference on Computer Vision Workshop (ICCVW), 27–28 Oct 2019

31. Yadav, J., Rani, A., Singh, V., Murari, B.M.: Prospects and limitations of non-invasive blood glucose monitoring using near-infrared spectroscopy. Biomed. Signal Process. Control **18**, 214–227 (2015)

32. Villena Gonzales, W., Mobashsher, A.T., Abbosh, A.: The progress of glucose monitoring—a review of invasive to minimally and non-invasive techniques, devices and sensors. Sensors **19**, 800 (2019)

33. Alsunaidi, B., Althobaiti, M., Tamal, M., Albaker, W., Al-Naib, I.: A review of non-invasive optical systems for continuous blood glucose monitoring. Sensors **21**, 6820 (2021)

34. Sun, L., Wang, L., Shen, W., Liu, C., Bai, F.: Robust RPPG method based on reference signal envelope to improve wave morphology. Electronics **12**, 2813 (2023)

35. Kim, S.-E., Yu, S.-G., Kim, N.H., Suh, K.H., Lee, E.C.: Restoration of remote PPG signal through correspondence with contact sensor signal. Sensors **21**, 5910 (2021)

36. Haugg, F., Elgendi, M., Menon, C.: Effectiveness of remote PPG construction methods: a preliminary analysis. Bioengineering **9**, 485 (2022)

37. Haan, G.D., Jeanne, V.: Robust pulse rate from chrominance-based rPPG. IEEE Trans. Biomed. Eng. **60**, 2878–2886 (2013)

38. Wang, W., Brinker, A.C.D., Stuijk, S., Haan, G.D.: Algorithmic principles of remote PPG. IEEE Trans. Biomed. Eng. **64**, 1479–1491 (2017)

39. Pelletier, K.R.: Theory and applications of clinical biofeedback. J. Contemp. Psychother. **7**, 29–34 (1975)

40. Goessl, V.C., Curtiss, J.E., Hofmann, S.G.: The effect of heart rate variability biofeedback training on stress and anxiety: a meta-analysis. Psychol. Med. **47**, 2578–2586 (2017)

41. Nogueira, P.A., Urbano, J., Reis, L.P., Lopes Cardoso, H., Silva, D.C., Rocha, A.P.C., Gonçalves, J., Faria, B.M.: A review of commercial and medical-grade physiological monitoring devices for biofeedback-assisted quality of life improvement studies. J. Med. Syst. **42**, 1–10 (2018)
42. Can, Y.S., Iles-Smith, H., Chalabianloo, N., Ekiz, D., Fernández-Álvarez, J., Repetto, C., Riva, G., Ersoy, C.: How to relax in stressful situations: a smart stress reduction system. Healthcare **8**, 100 (2020)
43. Cohen, S., Janicki-Deverts, D., Miller, G.E.: Psychological stress and disease. JAMA **298**(14), 1685–1687 (2007)
44. Wirtz, P.H., Von Känel, R.: Psychological stress, inflammation, and coronary heart disease. Curr. Cardiol. Rep. **19**, 1–10 (2017)
45. Chong, C.S.-M., Tsunaka, M., Tsang, H.W.-H., Chan, E.P., Cheung, W.M.: Effects of yoga on stress management in healthy adults: a systematic review. Altern. Ther. Health Med. **17**(1), 32–38 (2011)
46. Khoury, B., Sharma, M.K., Rush, S.E., Fournier, C.-A.: Mindfulness-based stress reduction for healthy individuals: a meta-analysis. J. Psychosom. Res. **78**(6), 519–528 (2015)
47. Rudics, E., Nagy, Á., Dombi, J., Hompoth, E.A., Szabó, Z., Horváth, R., Balogh, M., Lovas, A., Bilicki, V., Szendi, I.: Photoplethysmograph based biofeedback for stress reduction under real-life conditions in healthcare frontline. Appl. Sci. (2023)
48. Van Der Zwan, J.E., De Vente, W., Huizink, A.C., Bögels, S.M., De Bruin, E.I.: Physical activity, mindfulness meditation, or heart rate variability biofeedback for stress reduction: a randomized controlled trial. Appl. Psychophysiol. Biofeedback **40**, 257–268 (2015)
49. Zhao, L., Zhang, X., Niu, X., Sun, J., Geng, R., Li, Q., Zhu, X., Dai, Z.: Remote photoplethysmography (rPPG) based learning fatigue detection. Appl. Intell. **53**, 27951–27965 (2023)
50. Brown, D.M.Y., Bray, S.R.: Heart rate biofeedback attenuates effects of mental fatigue on exercise performance. Psychol. Sport Exerc. **41**, 70–79 (2019)
51. Casado, C.Á., Cañellas, M.L., L'opez, M.B.: Depression recognition using remote photoplethysmography from facial videos. IEEE Trans. Affect. Comput. **14**, 3305–3316 (2022)
52. Pérez-Gaido, M., Lalanza, J.F., Parrado, E., Capdevila, L.: Can HRV biofeedback improve short-term effort recovery? Implications for intermittent load sports. Appl. Psychophysiol. Biofeedback **46**, 215–226 (2021)
53. Kantan, P.R., Spaich, E.G., Dahl, S.: A technical framework for musical biofeedback in stroke rehabilitation. IEEE Trans. Human-Mach. Syst. **52**, 220–231 (2020)